Superconductors: Experimental Aspects

Volume II

Superconductors: Experimental Aspects
Volume II

Edited by **Jared Jones**

New York

Published by NY Research Press,
23 West, 55th Street, Suite 816,
New York, NY 10019, USA
www.nyresearchpress.com

Superconductors: Experimental Aspects
Volume II
Edited by Jared Jones

International Standard Book Number: 978-1-63238-430-0 (Hardback)

Contents

Preface

The science of superconductivity is expanding speedily and new inventions are possible in the near future. This book is a compilation of researches accomplished by experts dealing with a variety of experimental aspects of superconductivity such as NMR studies, X-Ray spectroscopy studies, magnetic texturing, superconducting magnet technology, etc. The authors have attempted to express their unique vision and give an insight into the examined area of research. With this book, we intend to help our readers in understanding the above stated topic in a better way.

Various studies have approached the subject by analyzing it with a single perspective, but the present book provides diverse methodologies and techniques to address this field. This book contains theories and applications needed for understanding the subject from different perspectives. The aim is to keep the readers informed about the progress in the field; therefore, the contributions were carefully examined to compile novel researches by specialists from across the globe.

Indeed, the job of the editor is the most crucial and challenging in compiling all chapters into a single book. In the end, I would extend my sincere thanks to the chapter authors for their profound work. I am also thankful for the support provided by my family and colleagues during the compilation of this book.

Editor

Experiment

Field-Induced Superconductors: NMR Studies of λ–(BETS)$_2$FeCl$_4$

Guoqing Wu and W. Gilbert Clark

Additional information is available at the end of the chapter

1. Introduction

It has been widely known in condensed matter and materials physics that the application of magnetic field to a superconductor will generally destroy the superconductivity as a usual scenario. There are two reasons responsible for this.

One is the Zeeman effect [1-2], where the alignment of the electron spins by the applied magnetic field can break apart the electron pairs for a spin-singlet (but not a spin-triplet) state. In this case, the electron spin pairs have opposite spins (such as the s-wave spins typical in type I superconductors). The applied magnetic field attempts to align the spins of both electrons along the field, thus breaking them apart if strong enough. In terms of energy, one electron gains energy while the other (as a pair) loses energy. If the energy difference is larger than the amount of energy holding the electrons together, then they fly apart and thus the superconductivity disappears. But this does not apply to the spin triplet superconductivity (p-wave superconductors) where the electron pairs already have their spins aligned (along the field) in a p-wave state.

The other is the orbital effect [3], which is a manifestation of the Lorentz force from the applied magnetic field since the electrons (as a pair) have opposite linear momenta, one electron rotating around the other in their orbitals. The Lorentz force on them acts in opposite directions and is perpendicular to the applied magnetic field, thus always pulling the pair apart. This does not matter with their spin pairing symmetries (s-wave, p-wave or d-wave). In type-II superconductors, the Meissner screening currents associated with the vortex penetration in the applied magnetic field can also increase the electron kinetic energy (and momentum). Once this energy becomes greater than the energy that unites the two electrons, the electron pairs break apart and thus superconductivity is suppressed. Therefore, the orbital effect could be even more important in type-II superconductors.

However, in certain complex compounds, especially in some low-dimensional materials, superconductivity can be enhanced [4 - 5] by the application of magnetic field. The enhancement of superconductivity by magnetic field is a counter-intuitive unusual phenomenon.

In order to understand this interesting phenomenon, different theoretical mechanisms have been proposed, while there are still debates and experimental evidence is needed. The first theory is the Jaccarino-Peter compensation effect [6], the second theory is the suppression effect of the spin fluctuations [7-9], and the third theory is the anti-proximity effect (in contrary to the proximity effect [10]) found in the nanowires recently [11]. These theories will be briefly described in Section 3.

Experimentally, there are several effective techniques that can be used to the study of the superconductivity of the materials with magnetic field applications. They include electrical resistivity measurements, Nernst effect measurements, SQUID magnetic susceptibility measurements, and nuclear magnetic resonance (NMR) measurements, etc.

Among these experimental techniques, NMR is one of the most powerful ones and it is a versatile local probe capable of directly measuring the electron spin dynamics and distribution of internal magnetic field including their changes on the atomic scale. It has been widely used as a tool to investigate the charge and spin static and dynamic properties (including those of the nano particles). It is able to address a remarkably wide range of questions as well as testing the validity of existing and/or any proposed theories in condensed matter and materials physics.

The authors have extensive experience using the NMR and various other techniques for the study of the novel condensed matter materials. This chapter focuses on the NMR studies of the quasi-two dimensional field-induced superconductor λ–(BETS)$_2$FeCl$_4$. This is a chance to put some of the work together, with which it will help the science community for the understanding of the material as well as for the mechanism of superconductivity. It will also help materials scientists in search of new superconductors.

2. Field-induced superconductors

The discovery of field-induced superconductors is about a decade earlier than the discovery of the high-T_c superconductors (which have been found since 1986), while not many field-induced superconductors have been found. Both types of materials, field-induced superconductors and high-T_c superconductors, are highly valuable in science and engineering due to their important physics and wonderful potentials in technical applications.

Here are typical field-induced superconductors found so far, with chemical compositions as shown in the following

1. Eu$_x$Sn$_{1-x}$Mo$_6$S$_{8-y}$Se$_y$, Eu$_x$La$_{1-x}$Mo$_6$S$_8$, PbGd$_{0.2}$Mo$_6$S$_8$, etc.[5, 7].
2. λ–(BETS)$_2$FeCl$_4$, λ–(BETS)$_2$Fe$_x$Ga$_{1-x}$Br$_y$Cl$_{4-y}$, κ–(BETS)$_2$FeBr$_4$, etc.[4, 8, 9].
3. Al-nanowires (ANWs), Zinc-nanowires (ZNWs), MoGe nanowires, and Nb nanowires, etc. [10-13].

3. Theory for field-induced superconductors

In this section, we will briefly describe the theoretical aspects for the field-induced superconductors regarding their mechanisms of the field-induced superconductivity. We will mainly discuss the theory of Jaccarino-Peter effect, the theory of spin fluctuation effect, and the theory of anti-proximity effect.

3.1. Theory of Jaccarino-Peter effect

This theory was proposed by Jaccarino, V. and Peter, M. in 1962 [6]. It means that if there is an existence of localized magnetic moments at a state and conduction electrons as well at the same state in a material, then it could lead to a negative exchange interaction J between the conduction electrons and the magnetic moments when an external magnetic field (H) is applied. This negative exchange interaction J is formed due to the easy alignment of the localized magnetic moments (along the external magnetic field direction) while the external magnetic field H is applied. Thus the spins of the conduction electrons will experience an internal magnetic field (H_J), $H_J = J<S>/g\mu_B$, created by the magnetic moments [proportional to the average spins ($<S>$) of the moments]. Here the internal magnetic field H_J is called the exchange field and the direction of the exchange field H_J is opposite to the externally applied magnetic field H. This picture is sketched as that shown in Fig. 1.

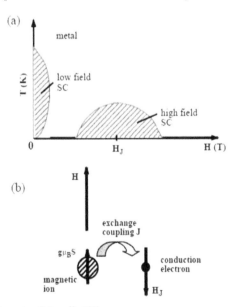

Figure 1. Schematic of Jaccarino-Peter effect [8]

In some cases, this H_J could be very strong. Therefore, if the exchange field H_J is strong enough to cancel the externally applied magnetic field H, i.e., $H_J = - H$, then the resultant field in total that the conduction electron spins experience becomes zero (also a complete suppression of the Zeeman effect), and thus the superconductivity is induced in this case.

Certainly, superconductivity could also be possible in this case (as a stable phase), even if without the external field H (i.e. $H = 0$). This is because the magnetic moments point to random directions (without H) and cancel each other, i.e., $H_J = 0$, and thus the conduction electron spins also feel zero in total field.

3.2. Theory of spin fluctuation effect

This theory was mainly reported by Maekawa, S. and Tachiki, M. [7] in 1970s, with the discovery of field-induced superconductors $Eu_xSn_{1-x}Mo_6S_{8-y}Se_y$. These types of materials have rare-earth 4f-ions and paired conduction electrons from the 4d-Mo-ions. The rare-earth 4f-ions have large fluctuating magnetic moments, while the conduction electrons from the 4d-Mo-ions have strong electron–electron interactions and they form Cooper pairs.

Without externally applied magnetic field ($H = 0$), the fluctuating magnetic field from the rare-earth 4f-ion moments at the 4d-Mo conduction electrons are so strong that it weakens the BCS coupling of the Mo-electrons. Thus there is no superconductivity without externally applied magnetic field.

However, when external magnetic field is applied ($H \neq 0$) it suppresses the spin fluctuation, causing an increase of the BCS coupling among the conduction electrons. Thus, superconductivity appears in the presence of an applied magnetic field. This scenario can be seen from a typical H-T phase diagram shown in Fig. 2.

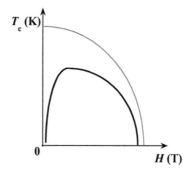

Figure 2. A possible H-T phase diagram for the spin fluctuation effect. The thin-red line represents a case without local spin fluctuation moments as a reference.

3.3. Theory of anti-proximity effect

3.3.1. Superconductivity in nanowires enhanced by applied magnetic field

Unlike the bulk superconductors, a nanoscale system can have externally applied magnetic field H to penetrate it with essentially no attenuation at all throughout the whole sample.

But it has been observed that the application of a small magnetic field H can decrease the resistance in even simple narrow superconducting wires (i.e., negative magnetoresistance) [12, 13], while larger applied magnetic field H can increase the critical current (I_c) significantly [14]. These indicate an enhancement of superconductivity in nanowires by the application of magnetic field. But to understand the enhancement of superconductivity by magnetic field in nanoscale systems is very challenging currently in the science community.

3.3.2. Proximity effect

On the other hand, when a superconducting nanowire is connected to two normal metal electrodes, generally a fraction of the wire is expected to be resistive, especially when the wire diameter is smaller than the superconducting coherence length. This is called the proximity effect [10].

Similarly, when a superconducting nanowire is connected to two bulk superconducting (BS) electrodes, the combined sandwiched system is expected to be superconducting (below the T_c of the superconducting nanowire and the BS electrodes), and the superconductivity of the nanowire is then expected to be more supportive and more robust through its coupling with the superconducting reservoirs. This is also actually what is theoretically expected [15].

3.3.3. Anti-proximity effect

Contrary to the proximity effect, it has been found in 2005 [11, 16] that, in a system consisting of 2 μ long, 40 nm diameter Zinc nanowires sandwiched between two BS electrodes (Sn or In), superconductivity of Zinc nanowires is completely suppressed (or partially suppressed) by the BS electrodes when the BS electrodes are in the superconducting state under zero applied magnetic field. However, when the BS electrodes are driven normal by an applied magnetic field (H), the Zinc nanowires re-enter their superconducting state at ~ 0.8 K, unexpectedly. This is called "anti-proximity effect".

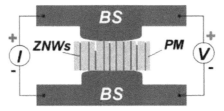

BS – bulk superconducting electrode; ZNWs – Zinc nanowires;
PM – porous membranes; I – current; V – voltage.

Figure 3. Schematic of the Zinc nanowires sandwiched between two BS electrodes [11].

This is also a counterintuitive unusual phenomenon, never reported before 2005.

The schematic of the electrical transport measurement system exhibiting the anti-proximity effect with the Zinc nanowires sandwiched between two BS electrodes is shown in Fig. 3.

3.3.4. Theory of anti-proximity effect

There are several theoretical models that could be used for the theoretical explanation of the magnetic field-induced or -enhanced superconductivity in nanowires, while some of which were proposed long before the anti-proximity effect was reported. Thus they are not generally accepted.

a.   Phase fluctuation model

This model proposes that there is an interplay between the superconducting phase fluctuations and dissipative quasiparticle channels [17].

The schematic diagram of this model regarding the anti-proximity effect experiment (Fig. 3) can be re-illustrated as that shown in Fig. 4.

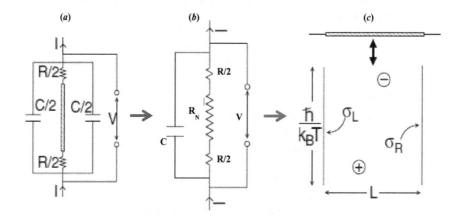

Figure 4. (a) Schematic of the anti-proximity effect experiment (Fig. 3). (b) Simplification of (a). (c) Phase fluctuations in a dissipative environment. R – resistance of the bulk electrodes,   C – circuit capacitance (between the electrodes),   V – voltage, I – current, σL and σR – superfluid (surface vortex densities) [17].

When the bulk electrodes are superconducting, there is a supercurrent flowing between the nanowire and BS electrodes, and the contact resistances (R) vanish (R = 0). Thus the circuit frequency becomes low, and the quantum wire is shunted by the capacitor (C) if the energy (frequency f) is less than the bulk superconducting gap energy of the electrodes. In this case, the quantum fluctuations of the superconducting phase drive the superfluid density (σ) of the zinc nanowire to zero (even when temperature $T = 0$). As soon as the superfluid density vanishes, the Cooper pairs dissociate and the nanowire becomes normal (with a normal resistor with resistance R_N). Thus this explains the resistive behavior at zero applied magnetic field ($H = 0$).

When the bulk electrodes are driven normal by the applied magnetic field ($H \geq 30$ mT), the contact resistances R $\neq$ 0. Similarly, if the electrodes are normal but the nanowire is superconducting (or vice versa), there will be a resistance due to charge conversion processes [10]. This results in a high circuit frequency $f = 1/(2\pi RC) \sim 1$ GHz, with a behavior like a pure resistor (i.e., impedance $X_C = 1/2\pi fC \rightarrow 0$). If this shunting resistance is less the quantum of resistance ($h/4e^2 \sim 6.4$ kΩ), then it will damp the superconducting phase fluctuations, and thus stabilizing the superconductivity. On the other hand, in a dissipation environment [Fig. 4 (c)], the superfluid density (σ) of the zinc nanowire cannot screen the interaction between bulk vortices completely. As a result, the superconducting phase becomes stable for sufficiently small shunt resistance [17].

In order words, when the magnetic field is applied ($H \neq 0$) to the bulk electrodes, the dissipations between the two ends of the electrodes will be enhanced and meanwhile the superconducting phase fluctuations are damped. This leads to the stabilization of the superconductivity of the nanowires between the bulk electrodes.

b. Interference model

The interference model proposes that there is an interference between junctions of two superconducting grains, with random Josephson couplings J and J' associated with disorder, as sketched shown in Fig. 5. It produces a configuration-averaged critical current <I_C> as [18]

$$\left\langle I_C \right\rangle = \left(J^2 + J'^2\right)^{1/2} \left[1 - \frac{1}{2}\left\langle \frac{JJ'}{J^2 + J'^2} \right\rangle \cos^2(2\pi \frac{\Phi}{\Phi_0}) + ...\right] \tag{1}$$

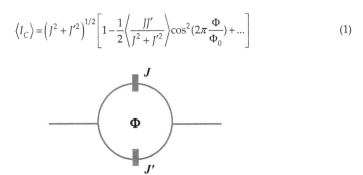

Here Φ represents for the magnetic flux through an array of each holes due to existence of disorder.

Figure 5. Schematic of interference between junctions with Josephson couplings J and J'

This is a periodic function of Φ [$\cos^2(2\pi\Phi/\Phi_0)$] with a period of half flux quantum ($\Phi_0/2 = hc/4e$), where Φ is the magnetic flux through each hole due to the existence of disorder in the sample (note, the sample has an array of holes through each of which there has a flux Φ).

Thus when Φ is small, <I_C> increases as Φ increases, and this corresponds to a negative magnetoresistance, i.e., when applied magnetic field H increases, the electrical resistivity of the nanowires drops down. Thereby the superconductivity is enhanced.

c.   Charge imbalance length model

This model proposes that there is a charge-imbalance length (or relaxation time) associated with the normal metal - superconductor boundaries of phase-slip centers [20]. Applying magnetic field reduces the charge-imbalance length (or relaxation time), resulting in a negative magnetoresistance at high currents and near T_c. Thus the superconductivity in the nanowires is enhanced.

d.   Impurity model

The impurity model deals with the superconductivity for nanoscale systems that have impurity magnetic moments with localized spins as magnetic superconductors [14], in which there is a strong Zeeman effect. According to this model, superconductivity is enhanced with the quenching of pair-breaking magnetic spin fluctuations by the applied magnetic field.

These are major theoretical models for the explanation of the anti-proximity effect in nanoscale systems. Their validity needs more experimental evidence.

4. Field-induced superconductor λ–(BETS)₂FeCl₄

The field-induced superconductor λ–$(BETS)_2FeCl_4$ is a quasi-two dimensional (2D) triclinic salt (space group P   ) incorporating large magnetic 3d-Fe^{3+} ions (spin S_d = 5/2) with the BETS-molecules inside which have highly correlated conduction electrons (π-electrons, spin S_π = 1/2) from the Se-ions, where BETS is bis(ethylenedithio)tetraselenafulvalene ($C_{10}S_4Se_4H_8$). It was first synthesized in 1993 by Kobayashi et al. [4, 8, 9, 19].

λ–$(BETS)_2FeCl_4$ is one of the most attractive materials in the last two decades for the observation of interplay of superconductivity and magnetism and for the synthesis of magnetic conductors and superconductors.

We expect it to show strong competition between the antiferromagnetic (AF) order of the Fe^{3+} magnetic moments and the superconductivity of the material, where the properties of the conduction electrons are significantly tunable by the external magnetic field, together with the internal magnetic field generated by the local magnetic moments from the Fe^{3+} ions as well. Thus it has been of considerable interest in condensed matter and materials physics.

This interplay originates from the role of the magnetic 3d-Fe^{3+} ions moments including the effect of their strong interaction with the conduction π-electrons. Because of this interplay, λ–$(BETS)_2FeCl_4$ has an unusual phase diagram [Fig. 5 (c)], including an antiferromagnetic insulating (AFI) phase, a paramagnetic metallic (PM) phase, and a field-induced superconducting (FISC) phase [4, 8].

The crystal structure of λ–$(BETS)_2FeCl_4$ in a unit cell is shown in Fig. 6 (a) [20]. In each unit cell, there are four BETS molecules and two Fe^{3+} ions. The BETS molecules are stacked along the a and c axes to form a quasi-stacking fourfold structure.

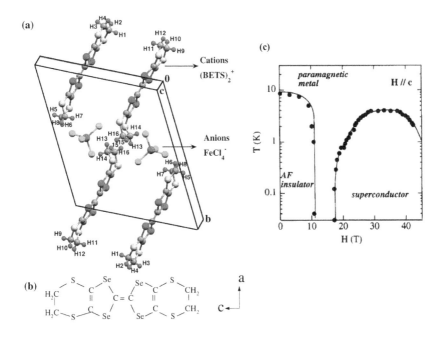

Figure 6. (a) Crystal structure of λ–(BETS)₂FeCl₄ in a unit cell. (b) BETS molecule [20]. (c) Phase diagram of λ–(BETS)₂FeCl₄ [8].

Noticeably, the conducting layers comprised of BETS are sandwiched along the b axis by the insulating layers of $FeCl^{4-}$ anions. The least conducting axis is b, the conducting plane is ac, and the easy axis of the antiferromagnetic spin structure is ~30° away from the c axis (parallel to the needle axis of the crystal) [21, 22].

At the room temperature (298 K), the lattice constants are: a = 16.164(3), b = 18.538(3), c = 6.592(4) Angstrom (A°), α = 98.40(1)°, β = 96.69(1)°, and γ = 112.52(1)°. The shortest distance between Fe^{3+} ions is 10.1 A° within a unit cell, which is along the a-direction, and the nearest distance of Fe^{3+} ions between neighboring unit cells is 8.8 A° [21].

5. NMR studies of λ–(BETS)₂FeCl₄

In order to study the mechanism of the superconductivity in λ–(BETS)₂FeCl₄ and to test the validity of the Jaccarino-Peter effect, as well as to understand the multi-phase properties of the material as show in the unusual phase diagram [Fig. 6 (c)], we successfully conducted a series of nuclear magnetic resonance (NMR) experiments.

These include both ⁷⁷Se-NMR measurements and proton (¹H) NMR measurements, as a function of temperature, magnetic field and angle of alignment of the magnetic field [20, 23, 24].

5.1. ^{77}Se-NMR measurements in λ–(BETS)$_2$FeCl$_4$

5.1.1. ^{77}Se-NMR spectrum

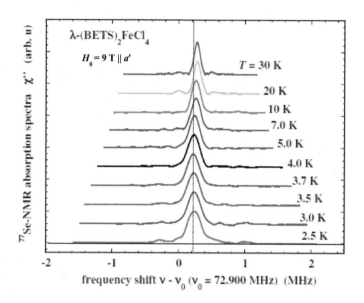

Figure 7. ^{77}Se-NMR absorption spectrum at various temperatures with applied magnetic field

$H_0 = 9$ T | | a' in λ–(BETS)$_2$FeCl$_4$ [23].

The ^{77}Se-NMR spectra of λ–(BETS)$_2$FeCl$_4$ at various temperatures are shown in Fig. 7. The spectrum has a dominant single-peak feature which is reasonable as a spin $I = 1/2$ nucleus for the ^{77}Se, while it broadens inhomogeneously and significantly upon cooling (the linewidth increases from 90 kHz to 200 kHz as temperature is lowered from 30 K to 5 K). What the ^{77}Se-NMR spectrum measures is the local field distribution in total at the Se sites. Apparently, these spectrum data indicate that all the Se sites in the unit cell are essentially identical.

The sample used for the ^{77}Se-NMR measurements was grown using a standard method [22] without ^{77}Se enrichment (the natural abundance of ^{77}Se is 7.5%). The sample dimension is $a^* \times b^* \times c = 0.09$ mm $\times 0.04$ mm $\times 0.80$ mm corresponding to a mass of ~ 7 μg with ~ 2.0×10^{15} ^{77}Se nuclei.

Due to the small number of spins, a small microcoil with a filling factor ~ 0.4 was used. For most acquisitions, 10^4–10^5 averages were used on a time scale of ~ 5 min for 10^4 averages. The sample and coil were rotated on a goniometer (rotation angle ϕ) whose rotation axis is along the lattice c-axis (the needle direction) and it is also perpendicular to the applied field

H_0 = 9 T. According to the crystal structure of λ–(BETS)₂FeCl₄, our calculation indicates that the direction of the Se-electron p_z orbital is 76.4° from the c-axis. Thus the minimum angle between p_z and H_0 during the rotation of the goniometer is ϕ_{min}=13.6°.

5.1.2. Temperature dependence of the ⁷⁷Se-NMR resonance frequency

The temperature (T) dependence of the ⁷⁷Se-NMR resonance frequency (v) from the above experiment is shown in Fig. 8 (a). In order to understand the origin of this resonance frequency, we also plotted it as a function of the 3d-Fe³⁺ ion magnetization (M_d), which is a Brillouin function of temperature T and the total magnetic field (H_T) at the Fe³⁺ ions. This is shown in Fig. 8 (b), where the solid lines show the fit to the M_d.

The resonance frequency v is counted from the center of the ⁷⁷Se-NMR spectrum peak (maximum). What it measures is the average of the local field in magnitude in total, including the direct hyperfine field from the conduction electrons and the indirect hyperfine field that coupled to the Fe³⁺ ions at the Lamar frequency of the ⁷⁷Se nuclei (see details in Section 5.1.4).

Figure 8 indicates that in the PM state above ~ 7 K at the applied field H_0 = 9 T, a good fit to frequency v (uncertainty ± 3 kHz) is obtained using

$$v(T,H_0) \approx a - bM_d(T,H_T), \tag{2}$$

where the fit parameters a = 73.221 MHz and b = 3.0158 [(mol.Fe/emu) .MHz].

This result is a strong indication that the temperature T dependence of the⁷⁷Se-NMR resonance frequency v is dominated by the hyperfine field from the Fe³⁺ ion magnetization M_d.

It is important to notice that the sign of the contribution from M_d is negative in Eq. (2). Thus, this also indicates that the hyperfine field from the Fe³⁺ ion magnetization is negative, i.e., opposite to applied magnetic field H_0, as needed for the Jaccarino-Peter compensation mechanism.

Now, to verify to validity of the Jaccarino-Peter mechanism, we need to find the field from the 3d Fe³⁺ ions at the Se π-electrons is (i.e., the π-d exchange field $H_{\pi d}$) which is the central goal of our ⁷⁷Se-NMR measurements.

According to the H-T phase diagram of λ–(BETS)₂FeCl₄ [Fig. 6 (c)], the magnitude of $H_{\pi d}$ = 33 T (tesla) at temperature T = 5 K.

5.1.3. Angular dependence of the ⁷⁷Se-NMR resonance frequency

The angular dependence of the ⁷⁷Se-NMR resonance frequency v from our experiments is shown in Fig. 9, which is plotted as a function of angle ϕ at several temperatures. The angle ϕ basically describes the alignment direction of the applied magnetic field H_0 relative to the sample lattice.

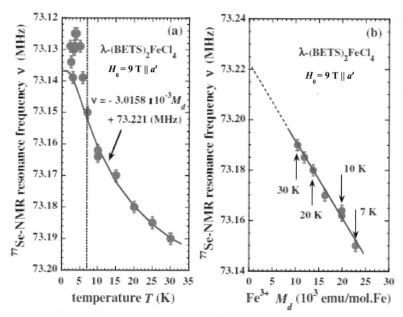

Figure 8. (a) ^{77}Se-NMR frequency shift as a function of temperature, and (b) ^{77}Se-NMR frequency shift vs the Fe^{3+} magnetization, with applied magnetic field $B_0 = 9$ T I I a' in λ–(BETS)₂FeCl₄ [23].

To understand the complexity of these sets of data, we need clarify the angle ϕ first as there are many other directions involved here as well. First, the crystal lattice has its a, b and c axes which have their own fixed directions. Second, the z component of the BETS molecule π-electron orbital moment, p_z, also has a fixed direction, which is perpendicular to the BETS molecule Se-C-S loop plane. Third, there is a direction of sample rotation which is along c (the needle direction) in the applied magnetic field H_0.

To distinguish each of these directions, we used the Cartesian xyz reference system and choose the reference z axis to be parallel to the lattice c axis, then the direction of p_z is determined to have angle 76.4° from the c axis through our calculation according to the X-ray data of λ–(BETS)₂FeCl₄. All these are clearly drawn as that shown in the inset of Fig. 9.

Thus during a sample rotation the direction of H_0 is always in the xy-plane, where x-axis is chosen to be in the c-p_z plane, and the angle ϕ is counted from the x-axis.

Therefore, an angle $\phi = 0°$ corresponds to H_0 to be in the c-p_z plane with the smallest angle between H_0 and p_z as $\phi_{min} = 13.6°$ as mentioned in Section 5.1.1.

Based on these data shown in Figs. 8 - 9, we can determine the magnitude of the π-d exchange field $H_{\pi d}$ precisely.

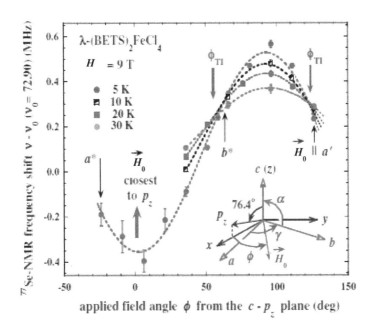

Figure 9. Angular dependence of the ^{77}Se-NMR resonance frequency v plotted at several temperatures for the rotation of H_0 = 9 T about the c axis in λ–(BETS)₂FeCl₄ [23].

5.1.4. Determination of the π-d exchange field
(between the Se-π and Fe^{3+}-d electrons)

From the theory of NMR [25], we can express the contributions to the Hamiltonian (H_I) of the ^{77}Se nuclear spins as

$$H_I \approx H_{IZ} + H_{I\pi}^{hf} + H_{Id}^{hf} + H_d^{dip},\tag{3}$$

where H_{IZ} is from the Zeeman contribution due the applied magnetic field, $H_{I\pi}^{hf}$ is from the direct hyperfine coupling of the ^{77}Se nucleus to the BETS π-electrons, while H_{Id}^{hf} is from the indirect hyperfine coupling via the π-electrons to the 3d Fe^{3+} ion spins, and the last term H_d^{dip} is from the dipolar coupling to the Fe^{3+} spins.

The π-d exchange field $H_{\pi d}$ comes from H_{Id}^{hf} , and the term H_d^{dip} produces dipolar field H_{dip}. We calculated H_{dip} from the summation of the near dipole, the bulk demagnetization and the Lorentz contributions. The cartoon of the π-d exchange interaction for the Jaccarino-Peter mechanism and the sample rotation direction in the magnetic field are shown in Fig. 10.

From Eq. (3) the corresponding ^{77}Se NMR resonance frequency ν is

$$
\begin{aligned}
\nu(\phi',H_0,T) \\
= {}^{77}\gamma\left[H_0 + H_{dip}(H_0,T)\right]\left[1 + K_c + K_s(\phi')\right] + {}^{77}\gamma\,K_s(\phi')H_{\pi d}(H_0,T),
\end{aligned}
\tag{4}
$$

where ϕ' is the angle between H_0 and the p_z directions, and K_c and $K_s(\phi')$ are, respectively, the chemical shift and the Knight shift of the BETS Se π-electrons.

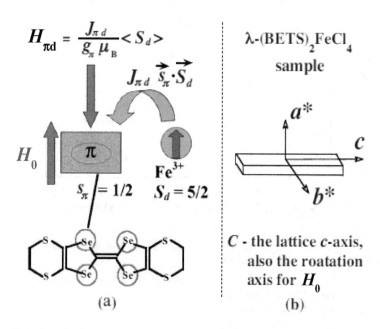

(a)   (b)

Figure 10. (a) Cartoon of the interactions for the Jaccarino-Peter mechanism. (b) Sketch of the sample rotation direction used for the ^{77}Se-NMR measurements [23].

Here,

$$
K_s(\phi') = K_{iso} + K_{an}(\phi') = K_{iso} + K_{ax}\left[3\cos^2\phi\cos^2\phi_{min} - 1\right],
\tag{5}
$$

where K_{iso} and K_{an} (ϕ') are the isotropic and axial (anisotropic) parts of the Knight shift, respectively. $K_{iso(ax)}$ is a constant determined by the isotropic (axial) hyperfine field produced by the $4p_\pi$ spin polarization of the BETS Se π-electrons [26].

The dashed lines in Fig. 8 are the fit to Eqs. (4) – (5). The gyromagnetic ratio of Se nucleus is $^{77}\gamma = 8.131$ MHz/T. The value of $K_{ax} = 15.3 \times 10^{-4}$ can be obtained precisely from the BETS molecule magnetic susceptibility and the π-electron spin polarization configuration [20, 26]. From the fit, now we can obtain the value of the π-d exchange field $H_{\pi d}$.

Alternatively, for better accuracy we obtained the following expression for the π-d exchange field $H_{\pi d}$ from Eqs. (4) – (5) to be,

$$H_{\pi d}(H_0, T_0) = \frac{\Delta \nu(T_0, \phi_1, \phi_2)}{^{77}\gamma \Delta K_{an}(\phi_1, \phi_2)} - \frac{H_{dip}(\phi_1) - H_{dip}(\phi_2)}{\Delta K_{an}(\phi_1, \phi_2)} - H_0. \tag{6}$$

Thus from the data T_0 = 5 K, ϕ_1 = 90°, and ϕ_2 = 0° as shown from Fig. 8, it gives $\Delta \nu(5K, 90°, 0°)$ = 880 ± 26 kHz, and $\Delta K_{an}(90°, 0°) = 4.42 \times 10^{-3}$. The value of $H_{dip}(90°) - H_{dip}(0°) = 3.63 \times 10^{-3}$ T is calculated with H_0 = 9.0006 T. With these values we obtained $H_{\pi d}$ = (- 32.7 ± 1.5) T at temperature T = 5 K and applied field H_0 = 9 T. This is very close to the expected value of −33 T obtained from the electrical resistivity measurement [27, 28] and the theoretical estimate [29].

If the applied field is H_0 = 33 T, by using our modified Brillouin function with the average of the Fe^{3+} spins, we expected the value of the $H_{\pi d}$ (33 T, 5 K) = (- 34.3 ± 2.4) T.

This large value of negative π-d exchange field felt by the Se conduction electrons obtained from our NMR measurements verifies the effectiveness of the Jaccarino-Peter compensation mechanism responsible for the magnetic-field-induced superconductivity in the quasi-2D superconductor λ–(BETS)₂FeCl₄.

6. Summary

We have presented briefly the information about the field-induced-superconductors including the theories explaining the mechanisms for the field-induced superconductivity. We also summarized our ^{77}Se-NMR studies in a single crystal of the field-induced superconductor λ–(BETS)₂FeCl₄, while most of our detailed research NMR work including both proton NMR and ^{77}Se-NMR were reported in refs.[20, 23, 24].

Our ^{77}Se-NMR experiments revealed large value of negative π-d exchange field ($H_{\pi d} \approx 33$ T at 5 K) from the negative exchange interaction between the large 3d-Fe^{3+} ions spins and BETS conduction electron spins existing in the material. This result directly verified the effectiveness of the Jaccarino-Peter compensation mechanism responsible for the magnetic-field-induced superconductivity in this quasi-2D superconductor λ–(BETS)₂FeCl₄.

Future high field NMR experiments ($H_0 \geq 30$ T) would be of interest, and NMR measurements with the alignment of applied magnetic field along the c-axis and sample rotation in the ac-plane (conducting plane) would further improve our understanding of this novel field-induced superconductor.

Author details

Guoqing Wu
Department of Physics, University of West Florida, USA

W. Gilbert Clark
Department of Physics and Astronomy, University of California, Los Angeles, USA

Acknowledgement

We are grateful to Prof. S. E. Brown at UCLA for helpful discussions and help with the NMR experiments. We also thank other coauthors J. S. Brooks, A. Kobayashi, and H. Kobayashi for our detailed ^{77}Se-NMR work published in ref. [23].

7. References

[1] Clogston, A. M., Upper limit for the critical field in hard superconductors, Phys. Rev. Lett. 15, 266 (1962).

[2] Chandrasekhar, B. S., A note on the maximum critical field of high-field superconductors. Appl. Phys. Lett. 1, 7 (1962).

[3] Tinkham, M., *Introduction to Superconductivity* (McGraw-Hill, New York, 1975).

[4] Uji, S., Shinagawa, S., Terashima, T., Yakabe, T., Terai, Y., Tokumoto, M., Kobayashi, A., Tanaka, H., Kobayashi, H., Magnetic-field-induced superconductivity in a two-dimensional organic conductor, Nature (London) 410, 908 (2001).

[5] Meul, H. W., Rossel, C., Decroux, M., et al., Observation of Magnetic-Field-Induced Superconductivity, Phys. Rev. Lett. 53, 497 (1984).

[6] Jaccarino, V. and Peter, M., Ultra-High-Field Superconductivity, Phys. Rev. Lett. 9, 290 (1962).

[7] Maekawa, S. and Tachiki, M., Superconductivity phase transition in rare-earth compounds. Phys. Rev. B 18, 4688 (1978).

[8] Uji, S., Kobayashi, H., Balicas, L. and Brooks, J. S., Superconductivity in an Organic Conductor Stabilized by a High Magnetic Field, Advanced Materials 14, 243 (2002).

[9] Kobayashi, H., Kobayashi, A., Cassoux, P., BETS as a source of molecular magnetic superconductors (BETS = bis(ethylenedithio)tetraselenafulvalene), Chem. Soc. Rev. 29, 325 (2000).

[10] Blonder, G. E., Tinkham, M. and K.lapwijk T. M., Transition from metallic to tunneling regimes in superconducting microconstrictions: Excess current, charge imbalance, and supercurrent conversion, Phys. Rev. B 25, 4515 (1982).

[11] Tian, M. L., Kumar, N., Xu, S. Y., Wang, J. G., Kurtz, J. S. and Chan, M. H. W., Phys. Rev. Lett. 95, 076802 (2005).

[12] Santhanam, P., Umbach, C. P. and Chi, C. C., Negative magnetoresistance in small superconducting loops and wires, Phys. Rev. B 40, 11392 (1989).

[13] Xiong, P., Herzog, A. V. and Dynes, R. C., Negative Magnetoresistance in Homogeneous Amorphous Superconducting Pb Wires, Phys. Rev. Lett. 78, 927 (1997).

[14] Rogachev, A., Wei, T.-C., Pekker, D., Bollinger, A. T., Goldbart, P. M. and Bezryadin, A., Magnetic-Field Enhancement of Superconductivity in Ultranarrow Wires, Phys. Rev. Lett. 97, 137001 (2006).

[15] Agassi, A. and Cullen, J. R., Current-phase relation in an intermediately coupled superconductor-superconductor junction, Phys. Rev.B, 54, 10112 (1996)

[16] Chen, Y. S., Snyder, S. D. and Goldman, A. M., Magnetic-Field-Induced Superconducting State in Zn Nanowires Driven in the Normal State by an Electric Current, Phys. Rev. Lett. 103, 127002 (2009).

[17] Fu, H. C., Seidel, A., Clarke, J. and Lee, D.-H., Stabilizing Superconductivity in Nanowires by Coupling to Dissipative Environments, Phys. Rev. Lett. 96, 157005 (2006); Sehmid, A., Diffusion and Localization in a Dissipative Quantum System, Phys. Rev. Lett. 51, 1506 (1983).

[18] Kivelson, S. A. and Spivak, B. Z., Aharonov-Bohm oscillations with period hc /4e and negative magnetoresistance in dirty superconductors, Phys. Rev. B 45, 10490 (1992).

[19] Kobayashi, A., Udagawa, T., Tomita, H., Naito, T., Kobayashi, H., New organic metals based on BETS compounds with MX_4^- anions (BETS = bis(ethylenedithio) tetraselenafulvalene; M = Ga, Fe, In; X = Cl, Br), Chem. Lett. 22, 2179 (1993).

[20] Wu, Guoqing, Ranin, P., Clark, W. G., Brown, S. E., Balicas, L. and Montgomery, L. K., Proton NMR measurements of the local magnetic field in the paramagnetic metal and antiferromagnetic insulator phases of λ–(BETS)₂FeCl₄, Physical Review B 74, 064428 (2006).

[21] Kobayashi, H., Fujiwara, F., Fujiwara, H., Tanaka, H., Akutsu, H., Tamura, I., Otsuka, T., Kobayashi, A., Tokumoto, M. and Cassoux, P., Development and physical properties of magnetic organic superconductors based on BETS molecules [BETS=Bis(ethylenedithio)tetraselenafulvalene], J. Phys. Chem. Solids 63, 1235 (2002).

[22] Kobayashi, H., Tomita, H., Naito, T., Kobayashi, A., Sakai, F., Watanabe, T. and Cassoux, P., New BETS Conductors with Magnetic Anions (BETS) bis(ethylenedithio)tetraselenafulvalene), J. Am. Chem. Soc. 118, 368 (1996).

[23] Wu, Guoqing, Clark, W. G., Brown, S. E., Brooks, J. S., Kobayashi, A. and Kobayashi, H., ⁷⁷Se-NMR measurements of the π-d exchange field in the organic superconductor λ–(BETS)₂FeCl₄, Phys. Rev. B 76, 132510 (2007).

[24] Wu, Guoqing, Ranin, P., Gaidos, G., Clark, W. G., Brown, S. E., Balicas, L. and Montgomery, L. K., ¹H-NMR spin-echo measurements of the spin dynamic properties in lambda-(BETS)₂FeCl₄, Phys. Rev. B 75, 174416 (2007).

[25] Slichter, C. P., *Principles of Magnetic Resonance*, 3rd ed. (Springer, Berlin, 1989).

[26] Takagi, S., *et al.*, ⁷⁷Se NMR Evidence for the Development of Antiferro-magnetic Spin Fluctuations of π-Electrons in λ-(BETS)₂GaCl₄, J. Phys. Soc. Jpn. 72, 483 (2003).

[27] Balicas, L., *et al.*, Superconductivity in an Organic Insulator at Very High Magnetic Fields, Phys. Rev. Lett. 87, 067002 (2001).

[28] Balicas, L., *et al.*, Pressure-induced enhancement of the transition temperature of the magnetic-field-induced superconducting state in λ-(BETS)₂FeCl₄, Phys. Rev. B 70, 092508 (2004).

[29] Mori, T. and Katsuhara, Estimation of πd-Interactions in Organic Conductors Including Magnetic Anions, M., J. Phys. Soc. Jpn. 71, 826 (2002).

Chapter 2

Defect Structure Versus Superconductivity in MeB2 Compounds (Me = Refractory Metals) and One-Dimensional Superconductors

A.J.S. Machado, S.T. Renosto, C.A.M. dos Santos, L.M.S. Alves and Z. Fisk

Additional information is available at the end of the chapter

1. Introduction

More than 24,000 inorganic phases are known. Of these phases approximately 16,000 are binary or pseudobinary while about 8,000 are ternary or pseudo-ternary. However, it is surprising to note that the observation of superconductivity in these alloys is a rare phenomenon. Superconductivity is ubiquitous but sparsely distributed and can be considered a rare phenomenon among the known alloys. BCS theory has been enormously successful in explaining the superconducting phenomena from the microscopic view point. The fundamental idea of this theory is the formation of Cooper pairs of electrons, mediated by phonons, the quantum of vibration of the crystal lattice [1]. Thus maximizing the critical temperature is involved with maximizing the electron-phonons coupling. Among the intermetallic materials, the binary cubic (A_3B) so-called A15 compounds displayed the highest T_c, until the discovery of superconducting cuprates. Among these materials in particular, Nb_3Sn and V_3Si with critical temperatures of 18.0 K and 17.1 K respectively have lattice instabilities of martensitic-type occurring at temperatures T_m very close to the maximum T_c. In the phase diagram of T_m and T_c versus Pressure (P) of V_3Si, the martensitic phase line intersects and stops exactly at the superconducting phase boundary. A qualitative example of this kind of the behavior can be observed in the Figure 1. Data exists beyond the extrapolated intersection shown and finds that there is no martensitic distortion occurring below T_c in this pressure regime. One way to think about this behavior is in terms of a lattice softening arising from strong electron-phonon coupling. Both the martensitic distortion and superconductivity arise from this coupling, and when the superconductivity T_c occurs at higher temperature than that of the lattice distortion, the energy gap that opens in the superconducting state gaps out at the same time phonon fluctuations that give rise to the lattice distortion [2-3]. One has, then, two phases that are competing for the same resource.

A similar type behavior is observed in heavy Fermion (HF) superconducting materials. Here antiferromagnetic order competes with the superconducting transition, both phases arising from electronic coupling to magnetic fluctuation in the heavy electron liquid.

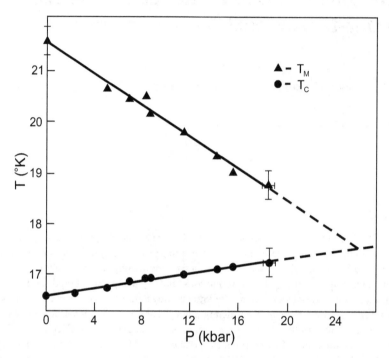

Figure 1. Schematic representation of the temperature dependence with pressure showing the critical temperature and martensitic temperature for V_3Si.

With the heavy Fermions, all superconductors are found in the vicinity of a quantum critical point, where the antiferromagnetic order has been driven to zero Kelvin (T = 0K). The general characteristics of high critical temperature cuprates are often discussed in terms of the kind of phase diagram found in the HF materials. A line known as the pseudogap intersects the maximum T_c in a superconducting dome in the temperature control parameter phase diagram, in general doping level being the control parameter [3]. There continues debate as to whether the pseudogap line represents a true phase transition. However, it is arguably the temperature setting the critical superconducting transition temperature upper limit. Again, the similarities with other instabilities discussed here are evident, with the temperature of the pseudogap intercepting the maximum of T_c against a control parameter that in this case is the doping level. This same discussion is also relevant to recent Fe-based pnictide superconductors. In this set of materials two competing phases are observed in the phase diagram, the non-superconducting one a structural instability or an SDW. Their transition temperatures intersect the superconducting transition temperature curve in the phase diagram versus pressure and/or composition. At the pressure suppressing the structural or SDW transition down to the superconducting T_c superconductivity does not appear to coexist with the SDW or structurally distorted phase. However, the data are not sufficient to say that the critical temperature of the superconducting transition is maximized at the intercept of the SDW transition, but the results seem to suggest that this may occur. In addition, the organic superconductors show similar behavior to that which occurs in high critical temperature cuprates. In all these cases maximizing the superconducting temperature appears to involve suppressing a secondary phase which competes with the phenomenon of superconductivity, whether CDW, SDW, T_m or other competitive instability. These experiments all suggest that superconducting pairs are utilizing the same fluctuation spectrum that supports the order of the phases it competes with. The key to superconductivity is to be found in frustrating the competing order so that the superconducting instability of the Fermi surface can gap out parts of the fluctuation spectrum favoring the second phase. Even in one-dimensional systems instabilities play a role in the superconducting behavior [4-5]. The superconducting critical temperature increases upon applying hydrostatic pressure while simultaneously suppressing the electronic CDW. Within this general scenario is the superconductivity of compounds which can crystallize in AlB2 prototype structure. Our discussion of superconductivity in this structure-type is based on the defect structures that these compounds may have.

1.1. Superconductivity in compounds with AlB₂ prototype structure

The first superconductor found to crystallize in the AlB₂ prototype structure was discovered by A. S. Cooper et. al. [6] In this paper the authors showed that NbB₂ as well MoB₂ can exhibit superconductivity. However, stoichiometric NbB₂ was not superconducting, but adding excess boron for a nominal composition of NbB₂.₅ yielded bulk superconductivity observed at 3.87 K, determined from the measurement of the specific heat. In the same article the authors discussed the high temperature phase MoB₂. When a nominal composition MoB₂.₅ was splat-melted forcing excess boron into lattice the material had a superconducting transition close to 7.45 K. These results were not confirmed by other authors and remained of little interest to the scientific community until the discovery of superconductivity in MgB₂ with critical temperature close to 40.0 K [7]. In the Nb-B system the NbB₂ phase shows a wide range of boron stoichiometry where a defect structure can explain this large range of solubility. Recently C. A. Nunes et. al. [8] made a systematic study of the B-solubility in NbB₂. The homogeneity of the phases obtained was determined by neutron diffraction and a maximum critical temperature was found to be 3.9 K. This study raises anew the question of the nature of defects that can be generated in these compounds. The stability of the phase NbB₂ spans a wide range of composition as shown in the diagram in Figure 2.

At the solubility limit the Nb-B inter-atomic distance in NbB₂ phase is constant. This indicates that variations in lattice parameters "a" and "c" inside of the stability range do not occur randomly but are such as to maintain a constant Nb-B distance of 2.43Å. To explain the wide stability range of the NbB_2 phase, some authors propose a superstructure of defect vacancies in the crystal lattice [10]. This defect structure is based on the cohesive forces in layers exerted by expansive forces in the Nb layers. Certainly this defect structure strongly influences the electronic structure, making it highly dependent on stoichiometry. The MoB_2 compound was also revisited for L. E. Muzzy et al. in [11] which reported a systematic study of the substitution of Mo by Zr in the $(Mo_{0.96}Zr_{0.04})_{0.88}B_2$ where the AlB_2 structure is stabilized and superconductivity can exist in the range of critical temperature between 5.9 and 8.2 K. In this case the authors claim that the superconductivity is strongly dependent on a specific defect structure such as occurs in NbB_2 compound. Another interesting compound that presents a relatively high superconducting critical temperature is $CaSi_2$. The equilibrium phase is hexagonal with space group R-3mh. However when this compound is submitted to the high pressure 15 GPa, an allotropic transformation occur into the AlB_2 prototype structure with sp^2 graphite-like planes, with superconducting critical temperature close to 14.0 K [12].

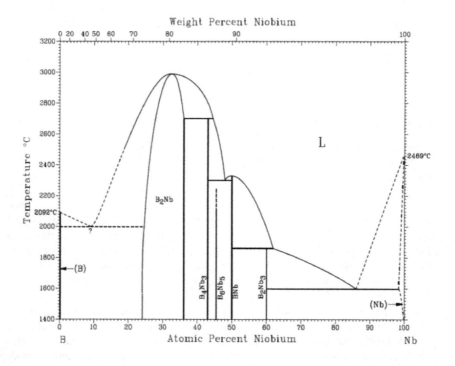

Figure 2. Phase diagram of the Nb – B system which show the wide solubility range of the NbB_2 phase. Adapted from reference [9].

Within this context we will discuss an example where an extremely stressed lattice of AlB_2 prototype yields superconductivity while the matrix compound is non-superconducting.

Defect Structure Versus Superconductivity in MeB2 Compounds (Me = Refractory Metals) and
One-Dimensional Superconductors

25

1.1.1. Superconductivity in a supersaturated solid solution of $Zr_{1-x}V_xB_2$

Since the discovery of superconductivity in MgB_2 with superconducting critical temperature close to 40 K, MB_2 materials (M = Transition Metal) with the same prototype structure as MgB_2 are considered as candidates for multiband superconductivity. As mentioned above, superconductivity in this class of material is relatively rare. Theoretical articles in the literature suggest that some member compounds are good candidates to exhibit high superconducting critical temperature. Among the suggested compounds are AuB_2, AgB_2, LiB_2, ZnB_2 and CaB_2 [13-17]. However, these theoretical predictions have not been confirmed and some of the suggested compounds do not exist in the equilibrium phase diagram. As example we mention ZnB_2 that does not exist in the Zn-B binary system in thermodynamic equilibrium. In the binary system (Zn-B) system no solubility exists between Zn and boron atoms and no intermetallic phase is observed. In our opinion the most important criterion for superconductivity in the AlB_2-type structure is defect creation such as occurs in NbB_2. Superconductivity was reported in ZrB_2 with superconducting critical temperature close to 5.5 K [18]. However, this surprising result was not confirmed by other groups. In fact single crystals of this compound (ZrB_2) are not superconducting [19]. This apparent paradox can be attributed to the possible existence of ZrB_{12} as contaminante in the sample prepared by Gasparov et. al. [18]. A careful analysis of the Zr-B phase diagram suggests that this kind of the contamination is quite possible, ZrB_{12} is a superconductor known for a long time, with superconducting critical temperature 5.94 K [20]. ZrB_2 also was studied at high pressure (50 Gpa) and superconductivity was not observed [21].

In order to address this problem we made a systematic study of ZrB_2 where Zr is substituted for V in VB_2 and V for Zr in ZrB_2. In both the Zr-B and V-B systems the AlB_2 structure exists and a solid solution is possible. We prepared $Zr_{1-x}V_xB_2$ compositions with x for $0.01 \leq x \leq 0.1$. All the samples were prepared by arc-melting together the high purity elements taken in the appropriate amounts in a Ti gettered arc furnace on a water-cooled Cu hearth under high purity argon. The samples were remelted five times to ensure good homogeneity. Due to the low vapor pressure of these constituent elements at their melting temperatures, the weight losses during arc melting were negligible (< 0.5%). The samples were characterized by x-ray diffraction with CuK_α radiation. The results of x-ray diffractometry are show in Figure 3.

All peaks can be indexed as belonging to AlB_2 prototype structure until x = 0.05 V content. For composition with x > 0.05, a segregation of secondary phase is observed, this secondary phase being interpreted as VB. These results indicate that the solubility limit is low. The lattice parameter as a function of V content is shown in Figure 4.

A small but consistent variation is observed indicating that the "c" lattice parameter systematically decreases with the substitution of Zr for V. However, the "a" lattice parameter is essentially constant as a function of V content. These results are consistent with the radius of V relative to Zr and indicates that V occupies the positions (0,0,0). The substitution of Zr for V yields a contraction in the Zr layers in the AlB_2 prototype structure. Indeed VB_2 presents a ~ 3.0 Å and c ~ 3.05 Å, while ZrB_2 has a ~ 3.16 Å and c ~ 3.53 Å. Thus the difference between both lattices parameters is about 5% relative to the "a" parameter and about 16% relative to the "c" parameter. These differences explain the larger variation of the "c" lattice parameter than "a" lattice parameter shown in Figure 4. The contraction in the crystalline structure occurs until x ~ 0.05 in the global composition indicating that the solubility limit is quite limited. This suggests that the integrity of unit cell strongly affects the electronic structure in this material and may change radically the electronic properties in the matrix compound (ZrB$_2$). The magnetization dependence with temperature is show in Figure 5 in which a superconducting transition emerges even at very low substitution of Zr by V at 6.4 K (x=0.01).

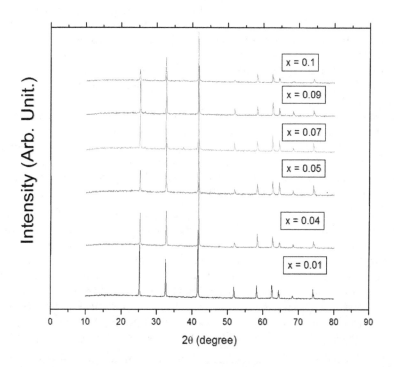

Figure 3. X-ray sequence of the samples with $0.01 \leq x \leq 0.1$ interval of composition in $Zr_{1-x}V_xB_2$.

In the inset the characteristic type II superconducting behavior is seen. These results are especially interesting because they suggest that the isoelectronic V can radically affect the electronic structure and is able to induce superconductivity in a non-superconducting matrix (ZrB$_2$). Indeed for the composition with x = 0.04 the superconducting critical temperature reaches the maximum value close to 8.52K. The dependence of magnetization with the temperature and applied magnetic field at 2.0K is shown in Figure 6. Once again clear superconducting behavior is observed close to 8.52K and the inset again reveals the type II superconducting behavior.

Defect Structure Versus Superconductivity in MeB2 Compounds (Me = Refractory Metals) and
One-Dimensional Superconductors

27

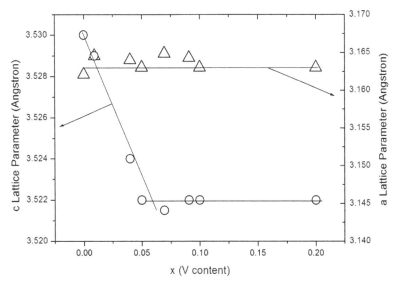

Figure 4. Lattice parameter variation as a function of vanadium content. The c lattice parameter
displays a small and consistent variation with vanadium content. The a lattice parameter is essentially
constant.

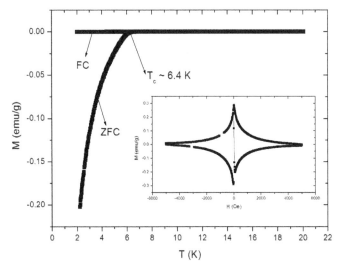

Figure 5. M vs T behavior which shows the critical temperature close to 6.4 K. The inset shows typical
type II superconducting behavior.

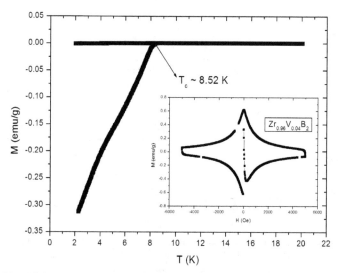

Figure 6. M vs T behavior which shows the critical temperature close to 8.52K. The inset shows typical type II superconducting behavior.

For compositions higher than x = 0.04 the critical temperature reaches a saturation value consistent with the solubility limit. For example, for a sample with composition $Zr_{0.95}V_{0.05}B_2$ the critical temperature is close to 8.2K as shown in Figure 7 where the dependence of the critical temperature as a function of V content is shown up to the solubility limit.

In samples with composition higher than x = 0.05 the critical temperature remains at 8.2K. Although the critical temperature does not change with higher V content, a decrease of the superconductor fraction is observed which is consistent with the appearance of a secondary phase.

This behavior is consistent with the solubility limit shown in Figure 4. The superconducting fraction estimate from figure 6 is about 45% of total volume of the sample indicating bulk superconductivity. The resistive behavior is shown in figure 8 where the superconducting critical temperature is consistent with the magnetization measurement shown in figure 6. The inset shows the resistivity behavior in applied magnetic field for $0 \leq \mu_0H \leq 6.0$ T. These results suggest that the upper critical field is very high since even with applied magnetic field of $\mu_0H = 6.0$ T the material is superconducting with critical temperature close to 6.9 K. Using the onset critical temperature it is possible to make an estimative of the upper critical field using the WHH formula [22] in the limit of short electronic mean-free path (dirty limit) given by:

$$\mu_0H_{c2}(0) = -0.693T_c \ (dH_{c2}/dT)_{T=Tc}. \tag{1}$$

Using this formula the upper critical field at zero Kelvin is estimated to be $\mu_0H_{c2}(0) \sim 17.9$ T, a surprisingly high upper critical field for this class of the material. Figure 9 shows the solid line expected in the WHH model, which fits the data very well and leads to the $\mu_0 H_{c2}(0)$ value of 17.9 T. Hence, pair breaking in this compound ($Zr_{0.96}V_{0.04}B_2$) is probably caused by orbital fields.

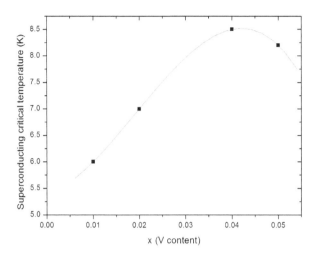

Figure 7. Superconducting critical temperature as a function of vanadium content.

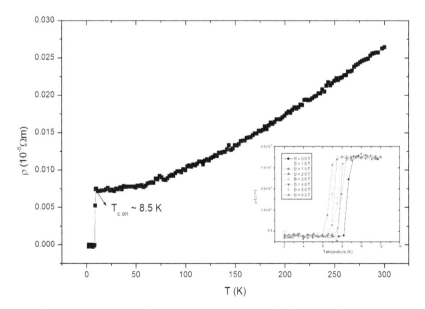

Figure 8. Resistivity as a function of temperature showing essentially the same critical temperature seen in M vs T. The inset shows the dependence of the critical temperature on applied magnetic field.

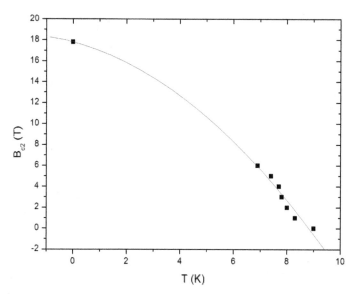

Figure 9. H_{c2} versus temperature extracted from resistivity measurements which show good agreement with WHH model (red line).

Bulk superconductivity is demonstrated by the heat capacity measurement shown in Figure 10. A clear jump is observed in C/T plotted against T^2 at zero magnetic field and the critical temperature is consistent in both resistivity and magnetization measurements. The normal state fit to the expression $C_n = \gamma T + \beta T^3$ by a least-squares analysis yields the values $\gamma = 3.8$ (mJ/molK^2) and $\beta = 0.034$ (mJ/molK^4). This result shows unambiguously that $Zr_{0.96}V_{0.04}B_2$ is a bulk superconductor. The subtraction of the phonon contribution allows us to evaluate the electronic contribution to the specific-heat, plotted as C_e/T vs T in the inset of figure 10.

An analysis of the jump yields $\Delta C_e/\gamma_n T_c \sim 0.49$ which is about 45% of the 1.43 value that is weak-coupling BCS prediction. This superconducting fraction is consistent with the estimative of fraction revealed by magnetization measurements displayed in Figure 6. Thus, these results show unambiguously that the substitution of Zr for V in the matrix ZrB_2 is able to induce bulk superconductivity in a matrix that is not a superconductor.

When this sample is annealed at 2000°C for 24 hours, the superconducting behavior disappears. This result strongly suggests that the original samples produced by arc-melting produce a supersaturated solution with V that does not exist in thermodynamic equilibrium. This supersaturation is close to a structural instability which probably is responsible for the superconducting behavior found in all as-cast samples. This interpretation provides an example of superconductivity that can emerge in the vicinity of some instability such as in A15 materials or other examples presented in introduction to this chapter.

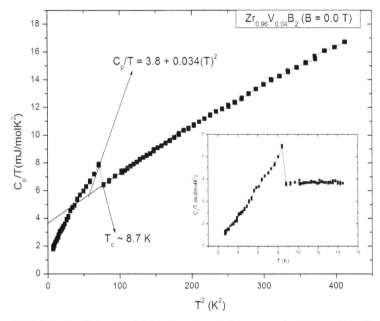

Figure 10. C_P/T against T^2 showing the clear jump close to the superconducting transition. The inset is shows the electronic contribution to the specific-heat.

1.1.2. Superconductivity in quasi-1D systems

During the last years great attention has been given to the study of superconductivity in low-dimensional (D) systems. This was motivated by the discovery of highly anisotropic behavior in high critical temperature cuprate and pnictide superconductors [22-24]. 1D systems are interesting because they are supposed to be simpler than 2D and 3D systems with regard to the electrical transport mechanisms. They offer a possibility to be compared with theoretical models for 1D conductors such as Luttinger Liquid (LL) theory [25,26] and Charge Density Wave (CDW) transition [6,7].

Generally, CDW transition exists in 1D systems as a consequence of the Peierls instability which is marked by the metal–insulator transition seen in electrical resistivity curves as a function of temperature [27]. One good example is the $K_{0.3}MoO_3$ compound [28]. On the other hand, LL behavior has been found only in special cases [29-30]. The best example recognized nowadays for the LL physics is the $Li_{0.9}Mo_6O_{17}$ purple bronze compound [31,32]. The electrical resistance behavior of this compound is well described by two power law temperature terms. The origin of the 1D electrical behavior of this compound is associated with Mo-O-Mo channels running along the b-axis of the monoclinic structure (for a good view of the crystalline structure see Fig. 2 of the reference [33]. The 1D behavior observed in the compound has been associated with the superconductivity with T_C = 1.9 K [34]. Furthermore, the 1D electrical conductivity, the charge hopping between 1D channels, the observation of Bose metal in the superconducting state, and the suppression of the metal-insulator transition at 25 K with increasing hydrostatic pressure along with consequent increasing of the superconducting critical temperature has been carefully discussed [31,34]. This has lead to a new discussion concerning whether the correlation between 1D behavior and superconductivity could exist in other compounds.

Based upon those observations in the purple bronze compound, our group has directed attention to the search for superconductivity in other molybdate oxides with low-D electrical conductivity. One example is the $A_xMoO_{2-\delta}$ compound with A = K or Na and x ≤ 0.3. Samples of this compound were prepared by solid state diffusion reaction using Mo, MoO_3, K_2MoO_4, and Na_2MoO_4 in appropriate amounts, encapsulated in quartz tube under vacuum or argon atmosphere, and heat treated at 400°C for 24 h followed by 700°C for 72 h. X-ray powder diffractometry showed that the samples are single phase and can be indexed with monoclinic structure of the space group $P2_1/c$ (14) and lattice parameters a = 5.62, b = 4.85, c = 5.63 Å , and β = 120.92°. The crystalline structure of the compound is shown in Fig. 12.

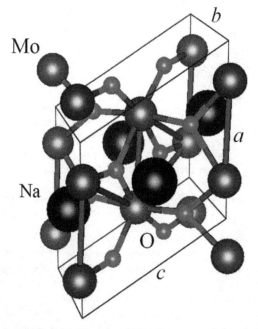

Figure 11. Crystalline structure of the $A_xMoO_{2-\delta}$ with A = Na or K.

1D channels of Mo-O-Mo run along a-axis which is responsible for the anisotropic electrical behavior of this compound. In fact, samples with A = Na and/or K show anomalous metallic behavior at low temperatures. Figure 13 displays the electrical resistance as a function of temperature for the $Na_{0.2}MoO_{2-\delta}$ samples.

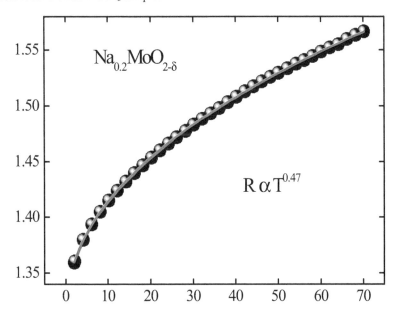

Figure 12. Typical anomalous electrical behavior observed in the $Na_{0.2}MoO_{2-\delta}$ compound.

A clearly anomalous non-linear behavior can be seen below 70 K. The electrical behavior can be well fit based upon the LL theory [4,5] using a single power law temperature term, $R \sim T^n$ with n = 0.468 ± 0.002 which is closed to the value expected in the LL theory (n = 0.5) [35]. The excellent quality of the fit suggests that the compound shows a quasi-1D electrical conductivity. All the samples studied in the $(Na,K)_x MoO_{2-\delta}$ system show similar power law temperature dependence below 70 K.

One of the most interesting aspects associated with this anomalous behavior is the existence of superconductivity in some samples. Figure 14 shows the magnetization as a function of temperature measured in the zero field cooled (ZFC) and field cooled (FC) condition in the $K_{0.05}MoO_{2-\delta}$ sample.

A superconducting temperature can be clearly observed at T_C = 4 K in the sample. Furthermore, a magnetic ordering is noticed at T_M = 65 K. The coexistence of the anomalous behavior, magnetic ordering, and superconductivity are common effects observed in several samples of $A_x MoO_{2-\delta}$ system. The correlation between the quasi-1D conductivity, superconductivity, and weak ferromagnetism is still under investigation [35-38].

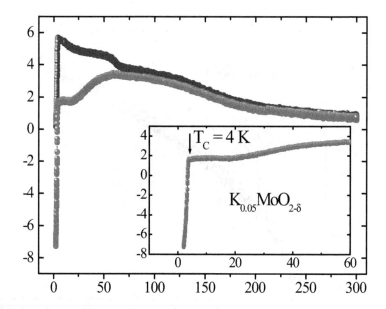

Figure 13. Magnetization as a function of temperature measured in the ZFC and FC procedures measured in the K$_{0.05}$MoO$_{2-\delta}$ sample.

2. Conclusions

This chapter reported the influence of the defects in the crystalline structure on the superconducting properties of intermetallic and low-dimensional compounds. The superconducting critical temperature of the materials can be associated with the electronic properties of the normal state. CDW, SDW, anisotropic behavior, structural phase transition, and doping content play important role on the superconducting properties of the compounds. Results about the influence of the crystalline structure on the superconducting properties for the MeB$_2$ and A$_x$MoO$_{2-\delta}$ compounds have been reported.

Author details

A.J.S. Machado, S.T. Renosto, C.A.M. dos Santos and L.M.S. Alves
Escola de Engenharia de Lorena, Universidade de São Paulo, Lorena, SP, Brazil

Z. Fisk
*Departments of Physics and Astronomy,
University of California at Irvine, Irvine, USA*

Defect Structure Versus Superconductivity in MeB2 Compounds (Me = Refractory Metals) and
One-Dimensional Superconductors

35

Acknowledgement

The authors are grateful by financial support through of the following grants (CNPq Brazilian Agency grant n° 303813/2008-3, 490182/2009-7 and 309084/2010-5) (Fapesp Brazilian Agency grant n° 2011/05961-3, 2009/54001-2, 2009/14524-6, 2010/06637-2 and 2010/11770-3) and AFOSR MURI.

3. References

[1] Bardeen, J.; Cooper, L. N.; Schrieffer, J. R. Phys. Rev., 108 (1957) 1175.

[2] C. W. Chu and L. R. Testardi, Phys. Rev. Lett., 32 (1974) 766.

[3] Z. Fisk, H. R. Ott and J. D. Thompson, Philosoph. Magazine, 89 (2009) 2111.

[4] C. Escribe-Filippini, J. Beille, M. Boujida, and C. Schlenker, Phys. C, 427 (1989)162.

[5] C. A. M. dos Santos, M. S. da Luz, Yi-Kuo Yu, J. J. Neumeier, J. Moreno, and B. D. White, Phys. Rev. B 77 (2008) 193106.

[6] A. S. Cooper, E. Corenzwit, L. D. Longinotti, B. T. Matthias and W. H. Zachariasen, Proceedings of the National Academy of Sciences, 67 (1970) 313.

[7] Nagamatsu J, Nakagawa N., Murakana Y. Z., Akimitsu J., Nature, 410, (2001), 63-64.

[8] Carlos Angelo Nunes, Dariusz Kaczorowski, Peter Rogle, Marica Regina Baldissera, Paulo Atsushi Suzuki, Gilberto Carvalho Coelho, Andriy Grytsiv, Gilles André, Francoise Bouree, Shigeru Okada, Acta Materialia, 53, (2005), 3679.

[9] T. Massalski. H. Okomono, P. Subramanian, L. Kacprozak (Eds), Binary Alloy Phase Diagrams, 2 and Edition American Society For Metals, Metals Park, 1990.

[10] Meerson G. A., Samsonov G. V., J. Chem. USSR, 27, (1954), 1053.

[11] L. E. Muzzy, M. Avdeev, G. Lawes, M . K. Haas, H. W. Zandbergen, A. P. Ramirez, J. D. Jorgensen, R. J. Cava, Physica C, 382 (2002) 153.

[12] S. Sanfilippo, H. Elsinger, M. Núñez-Regueiro and O. Laborde, Phys. Rev. B, 61 (2000) R3800.

[13] A. L. Ivanovskiĭ, Usp. Khim. 70 (9), 811 (2001).

[14] J. Kortus, I. I. Mazin, K. D. Belaschenko, *et al.*, Phys. Rev. Lett. 86 (20), 4656 (2001).

[15] J. M. An and W. E. Pickett, Phys. Rev. Lett. 86 (19), 4366 (2001).

[16] N. I. Medvedeva, A. L. Ivanovskii, J. E. Medvedeva, and A. J. Freeman, Phys. Rev. B 64, 020502 (2001).

[17] I. R. Shein, and A. L. Ivanovskii, Phys. Sol. State, vol. 44, N.10, 1833-1839 (2002).

[18] V. A. Gasparov, N. S. Sidorov, I. L. Zver'kova, and M. P. Kulakov, Pis'ma Zh. Éksp. Teor. Fiz. 73 (10), 601 (2001) [JETP Lett. 73, 532 (2001)].

[19] V. A. Gasparov, N. S. Sidorov and I. I. Zver'kova, Physical Review B, 73, 094510 (2006).

[20] Z. Fisk, A.C. Lawson, B.T. Matthias, E. Corenzwit, Phys. Lett. A 37, 251 (1971).

[21] A.S. Pereira, C.A. Perottoni, J.A.H. da Jornada, J.M. Leger, J. Haines, J. Phys. Cond. Mat., 14, 10615 (2002).

[22] N. R. Werthamer; E. Helfand; and P. C. Hohenberg, Phys. Rev., 147 (1966) 295.

[23] S. W. Tozer, A. W. Kleinsasser, T. Penney, D. Kaiser, and F. Holtzber, Phys. Rev. Lett. 59, 1768 (1987).

[24] M. A. Tanatar, N. Ni, C. Martin, R. T. Gordon, H. Kim, V. G. Kogan, G. D. Samolyuk, S. L. Bud'ko, P. C. Canfield, and R. Prozorov, Phys. Rev. B 79, 094507 (2009).

[25] V. N. Zavaritsky and A. S. Alexandrov, Phys. Rev. B 71, 012502 (2005).

[26] J. Voit, Rep. Prog. Phys. 58, 977 (1995).

[27] M. Ogata and P. W. Anderson, Phys. Rev. Lett. 70, 3087 (1993).

[28] W. L. McMillan, Phys. Rev. B 14, 1496 (1976).

[29] A. Berlinsky, Rep. Prog. Phys. 42, 1243 (1979).

[30] R. E. Thorne, Phys. Today, 42 (1996).

[31] R. Tarkiainen, M. Ahlskog, J. Penttilä, L. Roschier, P. Hakonen, M. Paalanen, and E. Sonin, Phys. Rev. B 64, 195412 (2001).

[32] S. N. Artemenko and S. V. Remizov, Phys. Rev. B 72, 125118 (2005).

[33] C. A. M. dos Santos, M. S. da Luz, Yi-Kuo Yu, J. J. Neumeier, J. Moreno, and B. D. White, Phys. Rev. B 77, 193106 (2008).

[34] C. A. M. dos Santos, B. D. White, Yi-Kuo Yu, J. J. Neumeier, and J. A. Souza, Phys. Rev. Lett. 98, 266405 (2007).

[35] M. S. da Luz, J. J. Neumeier, C. A. M. dos Santos, B. D. White, H. J. Izario Filho, J. B. Leão, and Q. Huang, Phys. Rev. B 84, 014108 (2011).

[36] C. Escribe-Filippini, J. Beille, M. Boujida, and C. Schlenker, Physica C 162-164, 427 (1989).

[37] L. M. S. Alves, V. I. Damasceno, C. A. M. dos Santos, A. D. Bortolozo, P. A. Suzuki, H. J. Izario Filho, A. J. S. Machado, and Z. Fisk, Phys. Rev. B 81, 174532 (2010).

[38] L. M. S. Alves, C. A. M. dos Santos, S. S. Benaion, A. J. S. Machado, B. S. de Lima, J. J. Neumeier, M. D. R. Marques, J. A. Aguiar, R. J. O. Mossanek, and M. Abbate. Submitted to Phys. Rev. B (2012).

Chapter 3

X-Ray Spectroscopy
Studies of Iron Chalcogenides

Chi Liang Chen and Chung-Li Dong

Additional information is available at the end of the chapter

1. Introduction

A recent study that identified high temperature superconductivity in Fe-based quatenary oxypnictides has generated a considerable amount of activity closely resembling the cuprate superconductivity discovered in the 1980s (Kamihare et al., 2008; Takahashi et al., 2008; Ren et al., 2008). This system is the first in which Fe plays an essential role in the occurrence of superconductivity. Fe generally has magnetic moments, tending to form an ordered magnetic state. Neutron-scattering experiments have demonstrated that mediated superconducting pairing may originate from magnetic fluctuations, similar to our understanding of that in high-T_c cuprates (de la Cruz et al., 2008; Xu et al., 2008). Binary superconductor $FeSe_x$ is another example of a Fe-based superconductor with a less toxic property, leading to the discovery of several superconducting compounds (Hsu et al., 2008). The T_c value of FeSe is ~8 K in bulk form and exhibits a compositional dependence such that T_c decreases for over-doping or under-doping of compounds (McQueen et al., 2009; Wu et al., 2009), as does that of high-T_c cuprates. FeSe has received a significant amount of attention owing to its simple tetragonal symmetry P4/nmm crystalline structure, comprising a stack of layers of edge-sharing $FeSe_4$ tetrahedron. The phase of FeSe heavily depends on Se deficiency and annealing temperature. While 400 °C annealing reduces the non-superconducting NiAs-type hexagonal phase and increases the PbO-type tetragonal superconducting phase (Hsu et al., McQueen et al., 2009; Wu et al., 2009; Mok et al., 2009), the role of Se deficiency remains unclear. Notably, this binary system is isostructural with the FeAs layer in quaternary iron arsenide. Also, band-structure calculations indicate that FeSe- and FeAs-based compounds have similar Fermi-surface structures (Ma et al., 2009), implying that this simple binary compound may significantly contribute to efforts to elucidate the origin of high-temperature superconductivity in these emerging Fe-based compounds. Therefore, although the electronic structure is of great importance in this respect, spectroscopic measurements are still limited.

According to investigations on how fluorine doping (Kamihara et al., 2008; Dong et al., 2008) and rare earth substitutions (Yang et al., 2009) influence the superconductivity in LaO$_{1-x}$F$_x$FeAs compounds, x-ray absorption spectroscopy (Kroll et al., 2008), x-ray photoemission spectroscopy (Malaeb et al., 2008) and resonant x-ray inelastic scattering (Yang et al., 2009) results, Fe 3d states hybridize with the As 4p states, leading to a situation in which itinerant charge carriers (electrons) are responsible for superconductivity. Most of these studies suggest moderate to weak correlate correlations in this system. Photoemission spectroscopy (PES) measurements (A. Yamasaki et al., 2010) support the density of state (DOS) calculations on the FeSe$_x$ system. These results indicate the Fe-Se hybridization and itinerancy with weak to moderate electronic correlations (Yoshida et al., 2009), while recent theoretical calculations have suggested strong correlations (Aichhorn et al., 2010; Pourret et al., 2011). While fluorine substitution leads to electron doping in the LaO$_{1-x}$F$_x$FeAs system, exactly how Se deficiency may bring in the mobile carriers in the FeSe$_x$ system to ultimately lead to superconductivity remains unclear. Therefore, this study elucidates the electronic structure of FeSe$_x$ (x=1~0.8) crystals by using XAS Fe and Se K-edge spectra. Powder x-ray diffraction (XRD) measurements confirm the lattice distortion. Analytical results further demonstrate a lattice distortion and Fe-Se hybridization, which are responsible for producing itinerant charge carriers in this system.

As mentioned earlier, although band-structure calculations indicate that FeSe and FeAs-based compounds have similar Fermi-surface structures, the poor quality of crystals arising from serious oxidization at their surfaces inhibit spectral measurements on pure (stoichiometric) FeSe. Also, FeSe exhibits an unstable crystalline structure. Therefore, investigating the effect of chemical substitution, at either the Se or Fe site, is a promising means of maintaining or improving the superconducting behavior on one hand and stabilizing the crystal structure on the other. Te doping of the layered FeSe with the PbO structure modifies its superconducting behavior, with a maximum T_c of ~ 15 K when Te replaces half of the Se. The improvement of T_c, which is correlated with the structural distortion that originates from Te substitution, is owing to the combined effect of lattice disorder, arising from the substitution of larger ions, and electronic interaction. Since layered FeSe$_{1-y}$Te$_y$ crystals are readily cleaved and highly crystalline, x-ray spectra of layered FeSe$_{1-y}$Te$_y$ crystals can provide clearer information about the electronic structure than those of FeSe crystals. Therefore, this study investigates the electronic properties of FeSe$_{1-y}$Te$_y$ (y=0~1) single crystals by using XAS and RIXS. XAS is a highly effective means of probing the crystal field and electronic interactions. The excitation-induced energy-loss features in RIXS can reflect the strength of the electron correlation. During their experimental and theoretical work on Fe-pnictides, Yang et al. (2009) addressed the issues regarding the Fe-based quaternary oxypnictides. However, few Fe-Se samples of this class have been investigated from a spectroscopic perspective. Angle-resolved photoemission (ARPES) combined with DFT band structure calculation on "11" Fe-based superconductor FeSe$_{0.42}$Te$_{0.58}$ reveals effective carrier mass enhancement, which is characteristic of a strongly electronic correlation (Tamai et al., 2010). This finding is supported by a large Sommerfeld coefficient, ~ 39 mJ/mol K (de la Cruz et al., 2008; Sales et al., 2009) from specific heat measurement. This phenomenon reveals that the FeSe "11" system is regarded as a strongly correlated system. Moreover, its electronic correlation differs markedly from that of "1111" and "122" compounds, perhaps due to the subtle differences between the p-d hybridizations in the Fe-pnictides and the FeSe "11" system. This postulation corresponds to the observation of p-d hybridization, as discussed later. This postulation is also supported by recent DMFT calculations, which demonstrate that correlations are overestimated largely owing to an incomplete understanding of the hybridization between the Fe d and pnictogen p states (Aichhorn et al., 2009). Nakayama et $al.$ (2010) discussed the pairing mechanism based on interband scattering, which has a signature of Fermi surface nesting in ARPES. Based on the SC gap value, their estimations suggest that the system is highly correlated (Nakayama et al., 2010). Moreover, a combined electron paramagnetic resonance (EPR) and NMR study of $FeSe_{0.42}Te_{0.58}$ superconductor has indicated the coexistence of electronic itinerant and localized states (Arčon et al., 2010). The coupling of the intrinsic state with localized character to itinerant electrons exhibits similarities with the Kondo effect, which is regarded as a typical interaction of a strongly correlated electron system. The localized state is characteristic of strong electron correlations and makes the FeSe "11" family a close relative of high-T_c superconductors. Comparing the XAS and RIXS spectra reveals that $FeSe_{1-y}Te_y$ is unlikely a weakly correlated system, thus differing from other Fe-based quaternary oxypnictides. The charge transfer between Se-Te and the Se $4p$ hole state induced by the substitution is strongly correlated with the superconducting behavior. Above results suggest strong electronic correlations in the FeSe "11" system, as discussed later in detail.

2. Experiments

$FeSe_x$ crystals were grown by a high temperature solution method described elsewhere (Wu et al., 2009; Mok et al., 2009). Crystals measuring 5 mm x 5 mm x 0.2 mm with (101) plate like habit could be obtained by this method. Three compositions results of $FeSe_x$ crystals (x=0.91, 0.88 and 0.85) are presented here for comparison and clarity. Additionally, large layered single crystals of high-quality $FeSe_{1-y}Te_y$ were grown using an optical zone-melting growth method. Elemental powders of $FeSe_{1-y}Te_y$ were loaded into a double quartz ampoule, which was evacuated and sealed. The ampoule was loaded into an optical floating-zone furnace, in which 2 x 1500 W halogen lamps were installed as infrared radiation sources. The ampoule moved at a rate of 1.5 mm/h. As-grown crystals were subsequently homogenized by annealing at 700 ~800 °C for 48 hours, and at 420 °C and for another 30 hours. Chemical compositions of $FeSe_{1-y}Te_y$ single crystals were determined by a Joel scanning electron microscope (SEM) coupled with an energy dispersive x-ray spectrometer (EDS) (Yeh et al., 2009; Yeh et al., 2008). In the Te substitution series, the composition of nominal y=0.3 was $FeSe_{0.56}Te_{0.41}$; that of nominal y=0.5 was $FeSe_{0.39}Te_{0.57}$; that of nominal y=0.7 was $FeSe_{0.25}Te_{0.72}$, and that of nominal y=1.0 was $FeTe_{0.91}$. The grown crystals were characterized by a Philips Xpert XRD system; T_c was confirmed by both transport and magnetic measurements (Wu et al., 2009; Mok et al., 2009).

X-ray absorption spectroscopy (XAS) provides insight into the symmetry of the unoccupied electronic states. The measurements at the Fe K-edge of chalcogenides were carried out at the 17C1 and 01C Wiggler beamlines at the National Synchrotron Radiation Reach Center

(NSRRC) in Taiwan, operated at 1.5GeV with a current of 360mA. Monochromators with Si (111) crystals were used on both the beam lines with an energy resolution $\Delta E/E$ higher than 2×10^{-4}. Absorption spectra were recorded by the fluorescence yield (FY) mode at room temperature by using a Lytel detector (Lytle et al., 1984). All spectra were normalized to a unity step height in the absorption coefficient from well below to well above the edges, subsequently yielding information of the unoccupied states with p character. Standard Fe and Se metal foils and oxide powders, SeO_2, FeO, Fe_2O_3 and Fe_3O_4 were used not only for energy calibration, but also for comparing different electronic valence states. Since surface oxidation was assumed to interfere with the interpretation of the spectra, the $FeSe_x$ crystals were cleaved *in situ* in a vacuum before recording the spectra.

The unoccupied partial density of states in the conduction band was probed using XAS, while information complementing that obtained by XAS was obtained using XES. Those results reveal the occupied partial density of states associated with the valence band. Detailed x-ray absorption and emission studies were conducted. Next, tuning the incident x-ray photon energies at resonance in XAS yields the RIXS spectrum, which is used primarily to probe the low-excited energy-loss feature which is symptomatic of the electron correlation. XAS and XES measurements of the Fe $L_{2,3}$-edges were taken at beamlines 7.0.1 and 8.0 at the Advanced Light Source (ALS) at Lawrence Berkeley National Laboratory (LBNL). In the Fe L-edge x-ray absorption process, the electron in the Fe $2p$ core level was excited to the empty $3d$ and $4s$ states and, then, the XES spectra were recorded as the signal associated partial densities of states with Fe $4s$ as well as Fe $3d$ character. The RIXS spectra were obtained by properly selecting various excitation energies to record the XES spectra, based on the x-ray absorption spectral profile. The XAS spectra were obtained with an energy resolution of 0.2 eV by recording the sample current. Additionally, the x-ray emission spectra were recorded using a high-resolution grazing-incidence grating spectrometer with a two-dimensional multi-channel plate detector with the resolution set to 0.6 eV (Norgdren et al., 1989). Surface oxidization is of priority concern in Fe-based superconductors; to prevent oxidation of the surface, all data were gathered on a surface of the sample that was cleaved *in situ* in a vacuum with a base pressure of 2.7×10^{-9} torr.

3. Results and discussion

3.1. Microstructure of FeSe and FeTe

Figure 1 (a) shows the tetragonal crystal structure of FeSe and its building blocks, i.e. Se-Fe tetrahedra and Fe-Se pyramidal sheets. Figure 1 (b) shows the electronic energy level of the individual constituent elements and $FeSe_x$ (indicating the hybridization band). Figure 2 shows the XRD patterns of the $FeSe_x$ (x=0.9, 0.88 and 0.85) crystals, which represent the superconducting FeSe phase. The patterns are correlated with the P4/nmm and indexed in the figure. Weak hexagonal reflections are also observed. According to this figure, the main diffraction peak (101) shown expanded in the inset shifts to a lower 2θ as x decreases, indicating an increase in the lattice parameters. The lattice parameters calculated from these patterns are a=b=3.771Å, c=5.528Å for x=0.9, a=b=3.775Å, c=5.528Å for x=0.88 and a=b=3.777Å,

c=5.529 Å for x=0.85. Experimental results indicate that the a=b parameter increases incrementally as x decreases. Simultaneously, a markedly smaller change occurs in the c-parameter. Thus, the a-b plane variation is surpasses that of the c axis. These lattice parameters are very close to those described in the literature for Se deficient powders (Hsu et al., 2008).

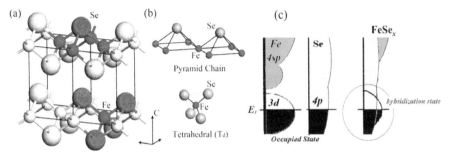

Figure 1. (a) Illustration of the crystal structure of tetragonal FeSe$_x$, where the blue (small balls) and red (large balls) denote Fe and Se, respectively; the pyramid chain and tetrahedral arrangements are marked in color in the unit cell. Hybridization is shown in two color bonds; (b) the local symmetry of Se atom (in pyramid chain) and Se atom (in tetrahedral geometry) shown separately; (c) energy level diagram of the FeSe$_x$ system along with individual elements. The hybridization and unoccupied states in the FeSe are highlighted by a circle.

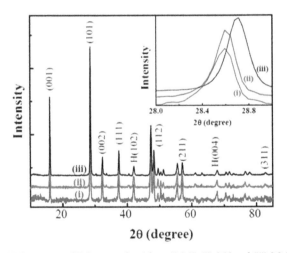

Figure 2. Powder XRD patterns of FeSe$_x$ crystals with x= (i) 0.85, (ii) 0.88 and (iii) 0.9. The patterns are fitted to the P4/nmm space group and indexed. Hexagonal phase reflections are denoted by a prefix H.

Conversely, FeTe with the same tetragonal crystal structure is stable up to a significantly higher temperature, ~1200 K. As is expected, replacing Se atoms within FeSe with Te stabilizes the tetragonal phase at a synthetic temperature close to or above 731 K. This observation correlates well with our X-ray diffraction analysis. This phenomenon is likely owing to that Te, which has a larger atomic size than Se, inhibits interatomic diffusion in the FeSe lattice. In contrast, Se atoms move easily in the larger FeTe lattice. The lattice parameters calculated from $FeSe_{1-y}Te_y$ patterns are increased with a increasing y (Yeh et al., 2009; Yeh et al., 2008). Analysis results indicate that T_c and gamma angle of the distorted lattice both reach a maximum value at $FeSe_{0.56}Te_{0.41}$ (~50% Te substituted). This correlation between the gamma angle and T_c indicates that T_c heavily depends on the level of lattice distortion and the distance of the Fe-Fe bond in the Fe-plane. Results of above studies correspond to the electron-orbital symmetry based on the XAS measurements, as discussed below.

3.2. Electronic structure results based on X-ray spectroscopy

3.2.1. FeSe$_x$ crystals

XAS spectra of the transition metal Fe K-edge ($1s \rightarrow 4p$) in Fig. 3(a) are mostly related to the partial density of $4p$ states of the iron site (Fig. 1(c)). Unoccupied states in the $3d$ (due to quadruple transition) and 4sp bands are sensitive to the local structure and the type of the nearest neighbors (de Groot et al., 2009; Chang et al., 2001; Longa et al., 1999). Above spectra can therefore obtain information about the changes in the electronic states possibly originating from changes in the environment of the Fe ions such as Se vacancies in the present case. Figure 3 shows the Fe K-edge absorption spectra of FeSe$_x$ crystals along with the standards Fe, FeO, Fe$_2$O$_3$ and Fe$_3$O$_4$. All spectra are normalized at the photon energy ~100 eV above the absorption edge at E_0 = 7,112 eV (the pure Fe K absorption edge energy). The spectra of the crystals appear to be close to the Fe metal foil, indicating that the crystals are free from oxidation. The recent Fe L-edge ($2p_{2/3} \rightarrow 3d$ transition) spectra measured before and after cleaving the samples in UHV further confirm this observation, as shown in the inset of Fig 3(a). The oxidation peak is observed in the crystal before cleaving but it does not appear after cleaving, indicating the possible formation of a thin oxide layer on the surface during handling. Such a thin layer may negligibly impact the deeper penetrating K-edge measurements. Since our measurements are made after cleaving the crystals under vacuum this possibility of surface oxidation is even eliminated. The following section describes in more detail the FeSe$_x$ electronic structure of the $3d$ states obtained by XAS measurements and resonant inelastic x-ray scattering (RIXS) at the Fe L$_{2,3}$ edges. The observations agree with those of Yang *et al.* (2009) on Fe-pnictides 1111 and 122 systems, as well as those of Lee *et al.* (2008) when using first principles methods to study FeSe$_x$ system. Therefore, our results on the oxygen free FeSe$_x$ crystals eliminate the possibility of oxygen in superconductivity, as is case in the $LaO_{1-x}F_xFeAs$ system.

These spectra reveal three prominent features, A$_1$, A$_2$ and A$_3$ (Fig. 3(a)), of which, A$_1$ could be assigned to the $3d$ unoccupied states originating from the Fe-Fe bonds in metallic iron. The features A$_2$ and A$_3$ refer to the unoccupied Fe 4sp states. The rising portion of the broad feature A$_2$ (~7,118.8 eV) appears as a broad peak labeled e1 at ~7,116.8 eV, which appears to be well separated in the first derivative plots of the spectra shown in Fig. 3(b). While not observed in the spectra of the reference Fe metal foil or oxides, this broad peak is a part of the Fe $4sp$ band. The e_1 feature appears at an energy between those of the Fe metal and FeO and, therefore, originates from a different interaction, as discussed later. Based on these first derivative plots, this study also evaluates a formal charge of Fe, in conjunction with the three standards FeO (Fe^{2+}), Fe_2O_3 (Fe^{3+}) and Fe_3O_4 ($Fe^{2.66+}$). Extrapolating the energy of FeSe0.88 with those of the standards allows us to obtain a formal charge of ~1.8+ for Fe in these crystals, thus establishing the electronic charge of Fe in the covalency (2+). Our results further indicate that the peak (energy) position does not increase in energy with a decreasing x, implying that the effective charge (valence) of Fe does not change with x. This finding is consistent with the above Fe L-edge spectra as well as RIXS analysis (Tamai et al., 2010). Thus, the possible electronic configurations of Fe in the ground state can be written as $3d^{6.2}$ or $3d^6 4s^{0.2}$, indicating a mixture of monovalency ($3d^6 4s^1$ or $3d^7$) and divalency ($3d^6$).

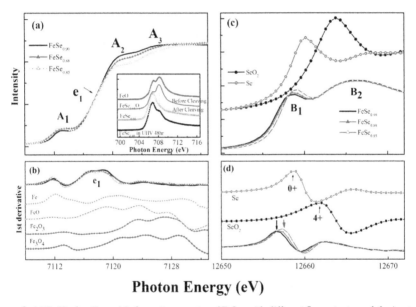

Figure 3. (a) Fe K-edge ($1s \rightarrow 4p$) absorption spectra of FeSe$_x$ with different Se contents, and the inset shows the XAS Fe L_3-edge spectra of FeSe0.88 crystal before and after cleaving *in situ* in a vacuum (b) the first derivative plots of the same spectra. The spectra of the standards - Fe-metal foil, FeO, Fe$_2$O$_3$ and Fe$_3$O$_4$ are also given alongside the sample spectra. (c)The Se K-edge ($1s \rightarrow 4p$) absorption spectra FeSe$_x$ with different Se contents, along with the standards Se metal and SeO$_2$ and (d) the first derivative plots of the same spectra.

Figure 3(c) shows the Se K-edge spectra of the FeSe$_x$ crystals and Se and SeO$_2$ standards, while Fig. 3(d) displays the corresponding first derivative plots to highlight energy changes in the spectra. The spectra represent mainly Se character without any trace of SeO$_2$, indicating the absence of oxidation, even in the deeper layers of the FeSe$_x$ crystals. The spectra exhibit two peaks B$_1$ and B$_2$. The B$_1$ feature at photon energy around 12,658 eV, formally assigned to the transitions $1s \rightarrow 4p$, increases slightly in intensity as well as shifts to a higher energy as x is decreased. This suggests an increase in the Se $4p$ unoccupied states i. e. in the upper Hubbard band (UHB). According to the first derivative plots, a formal charge of ~2.2- is obtained for Se in the x=0.88 crystal by interpolation with the energies of the standards Se and SeO_2 (as in the case of Fe). This finding agrees with a total charge of 0 when the formal charges of Fe and Se are added ($Fe^{1.8+}$ $Se_{0.88}^{2.2-}$), thus establishing the role of the electronic charge of Se in the covalency (2-).

The excess negative charge of -0.2 found in Se can be explained as follows. The electronic charge of Fe in the covalent $FeSe_x$ should be 2+. However, in this work, a formal charge of 1.8+ is obtained, implying that some electronic charge is returned to Fe due to the Se deficiency. However, the covalent charge of Se should be 2- yet 2.2- is obtained here, which is beyond the 6 electron occupancy of the $4p$ state orbitals. The Se K-edge spectra reveal a hole increase with a decreasing x; however, no change in the Fe K-edge suggests a change in valence. According to the soft x-ray photoemission spectroscopy (XPS) measurements of Yamasaki et al. (2010), the close distance between Se and Fe may increase the covalence of the Fe-Se bond due to the hybridization of Se $4p$ and Fe $3d$ states. Yoshida et al. (2009) also observed an adequate correlation between their DOS calculations and the XPS spectra, which show a feature corresponding to the Fe $3d$-Se $4p$ hybridization. Theoretical calculations of Subede et al. (2008) also suggest this Fe-Se hybridization. From the above discussion, we can infer that the B_1 feature (Fig. 3(c)) represents the Fe $3d$-Se $4p$ hybridization band (Fig. 1(c)). Additionally, the increased electronic charge on Fe mentioned above is due to itinerant electrons in the Fe-Se hybridization bond and appears as a hole increase in the Se K-edge spectra. Correlating this finding with the decreasing transition width in the resistance measurements (Mok et al., 2009) obviously reveals that the charge carriers responsible for superconductivity are itinerant electrons in a manner similar to the itinerant holes in the case of cuprates. Oxygen annealing in case of $YBa_2Cu_3O_{6+}$, oxidizes Cu^{2+} to Cu^{3+} through the hybridization of Cu $3d$-O $2p$ states. A previous study (Grioni et al., 1989; Merz et al., 1998) assigned the Cu^{3+} state to the empty state in the Cu-O bonds, which is also referred to as the $3d^9L$ ligand state, where a hole is located in the oxygen ions surrounding a 'Cu site' (L, a ligand hole, tentatively label this as $3d^8$-like). These itinerant holes are responsible for superconductivity. Similarly, in the case of $FeSe_x$, the Se deficiency is bringing about Fe $3d$-Se $4p$ hybridization leading to itinerant electrons. According to Mizuguchi et al. (2008), changes in bond lengths likely result in a reduction in the width of the resistive transition due to external pressure.

Experimental results indicate that the intensity of the A_2 feature diminishes as x is decreased, indicating a lattice distortion that increases with a decreasing x. Additionally, the change in the A_2 feature is larger than that of the A_3 feature. Notably, the A_2 feature could be associated with p_{xy} (σ) and A3 to p_z (π) orientations since multiple scattering in the XAS from p orbitals could reveal different orbital orientations, and also owing to the nature of the p orbitals, i.e. p_{xy} and p_z. We thus speculate that a larger distortion occurs in the a-b plane (Fe-Fe distance) than in the c axis. This is also seen from the XRD measurements (Fig. 2) where the change in the $a=b$ parameter is more profound than that of the c parameter (Fe-Se distance). Therefore, the Fe orbital structure changes from $4p$ to a varying (modulating) coordination with a decreasing x. The broad feature B$_2$, at ~ 20 eV above the Se K-edge appears at the same energy in the entire FeSe$_x$ spectra, is not affected by the Se deficiency and is assigned to the multiple scattering from the symmetrical Se $4p$ states in the coordination sphere that are correlated to the local structure of the Se ions (de Groot et al., 2009). This finding correlates with the XRD results where the c parameter remains nearly unchanged. This occurrence becomes obvious when examining Fig. 1, where Se is found at the tip of the Fe-Se pyramid.

Since Se is located at the apex of the tetrahedral pyramid chain in the FeSe$_x$ lattice (Fig. 1), removing a Se ion with a formal negative charge (-2) from the lattice would result in a Se vacancy with an effective positive charge and also repulse the surrounding positively charged Fe atoms. This finding corresponds to the distortion in the ab plane, as discussed above. Additionally, according to Lee *et al.* (2008), the Fe atoms around the vacancy may function similar to magnetic clusters.

3.2.2. Electron correlations of FeSe$_{1-y}$Te$_y$

As discussed above, due to the unstable phase of stoichiometry FeSe and efforts to more thoroughly understand the origin of superconductivity in this class of materials, of worthwhile interest is to investigate the effect of chemical substitution on FeSe in order to maintain or improve the superconducting property and stabilize the crystal structure. Figure 4(a) compares Fe $L_{2,3}$-edges XAS spectra of FeSe$_{1-y}$Te$_y$ before and after cleavage *in situ*, in which an oxidized iron foil is used as a reference sample. Of the two XAS lines, A$_1$ and B$_1$, at the L_3-edge of the oxidized iron foil, B$_1$ are prominent from the uncleaved samples, indicating that the sample surface is seriously contaminated by oxygen. The shoulder-like line B$_1$, which originates from Fe-O bonding in cleaved samples, is smeared. As is well known, iron oxide yields B$_1$. However, this line is missing from the spectra of cleaved FeSe$_{1-y}$Te$_y$ samples, indicating that iron oxide is only a very minor constituent. This finding reflects the importance of cleavage *in situ* for making spectral measurements of samples in this class. Figure 4(b) further illustrates how sensitive the FeSe (without Te doping) is, in which it oxidizes very easily, even in an ultra-high vacuum chamber with a base pressure of ~2.7×10^9 torr. The study of FeSe$_{0.88}$ crystal in an earlier section observed this phenomenon. In this work, XAS of pure FeSe is obtained, and the pure compound oxidizes within 12 hours, as indicated by a comparison with a doped sample (FeSe$_{1-y}$Te$_y$, y=0.5). The high-energy shoulder structure, as denoted by a dash line, is associated with the Fe-O bond; it is absent right after *in-situ* cleavage of either sample. This structure appears again in the FeSe sample when placed in a vacuum for 12 hr, yet does not reappear in the doped samples. This finding demonstrates that pure FeSe suffers from serious surface oxidation. Capable of enhancing the Tc value, Te doping can also stabilize the crystal structure. Figure 4(c) presents Fe $L_{2,3}$-edges XAS spectra of FeSe$_{1-y}$Te$_y$, by using Fe metal, oxidized foil and Fe$_2$O$_3$ as reference samples. Unoccupied Fe $3d$ orbitals are probed using XAS. The spectra reveal two major transitions. Governed by the dipole selection rule, the transition is mainly due to

$2p^63d^6$-$2p^53d^7$ transition, in which an Fe $2p$ electron is excited into an empty $3d$ state. Spin-orbit splitting separates the $2p$ state into $2p_{3/2}$ and $2p_{1/2}$ states, yielding two prominent absorption features around 707 eV (L_3) and 720 eV (L_2). XAS at the transition-metal L-edge probed unoccupied $3d$ states, which are sensitive to the chemical environment, valence state, crystal field and $3d$ electronic interactions. The ratio of intensities of the two major absorption features is largely determined by the high- or low-spin ground states through the crystal-field effect. XAS of Fe_2O_3 reveals a strongly split structure at both L_2 and L_3 absorption edges, which is formed by the interplay of crystal field and electronic interactions. Fe-O does not contribute to the $FeSe_{1-y}Te_y$ spectra, and no spectral profile resembles that of Fe_2O_3 or oxidized foil. As is generally assumed, the shoulder at the high-energy tail (dotted portion) of the Fe L_3 line in the $FeSe_{1-y}Te_y$ samples is associated with covalent sp^3 bonds between Fe $3d$ and Se $4p$/Te $5p$ states (Bondino et al., 2008). $FeSe_{1-y}Te_y$ yields no observable line splitting or change in intensity ratio, implying a weak crystal-field effect and also that the Fe ion favors a high-spin ground state. Line shapes in the spectra of these cleaved crystals in the UHV chamber resemble those of iron metal, indicating metallic nature and a localized $3d$ band (Yang et al., 2009). As is anticipated, the variation in the full width at half maximum (FWHM) of the Fe L_3 peak in the XAS and the width of the Fe $3d$ band with localized character is smaller than those of Fe because the Fe-Fe interaction is stronger (Nekrasov et al., 2008).

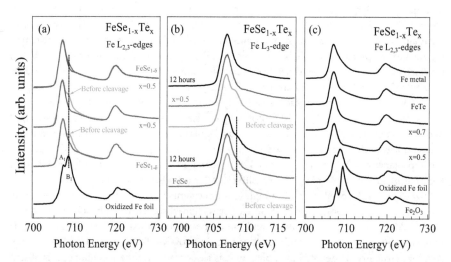

Figure 4. (a) Comparison of Fe $L_{2,3}$-edge XAS of $FeSe_{1-y}Te_y$ before and after cleavage *in situ*, using oxidized iron foil as a reference. (b) FeSe is highly sensitive, even in the vacuum; in addition, it oxidizes within 12 hours. (c) Fe $L_{2,3}$-edge XAS of $FeSe_{1-y}Te_y$ cleaved *in situ*. Samples of Fe metal, oxidized foil and Fe_2O_3 serve as references.

Figure 5(a) describes the RIXS (lower part) obtained at selected energies (letters a-g), as denoted by arrows in XAS (upper part). The upper RIXS spectrum is obtained at an excitation photon energy of 735 eV , far above the Fe L_3-edge absorption threshold. This spectrum is commonly referred to as the so-called non-resonant normal emission spectrum. This non-resonant spectrum reveals two main fluorescent features around 704 eV and 717 eV, resulting from de-excitation transitions from occupied $3d$ states to the $2p_{3/2}$ and $2p_{1/2}$ holes, respectively, which are thereby refilled. The RIXS recorded with various excitation energies, a-f, include the strong 704 eV peak, as observed in the non-resonant spectrum. At an excitation energy close to that of the L_2 absorption feature, the line at 717 eV emerges and remains at the same energy as the excitation energy increases beyond the L_2 edge. As indicated by arrows in spectra d and e, small bumps appear at energies of approximately 710 and 716 eV. These energies track well the excitation energies and, therefore, are caused by elastic scattering. As magnified in the inset, the fluorescent feature of both samples does not follow the excitation energy and well overlaps the line at 704 eV in the non-resonant spectrum. Notably, RIXS includes no energy loss feature, which is generally associated with electron correlation and excitations.

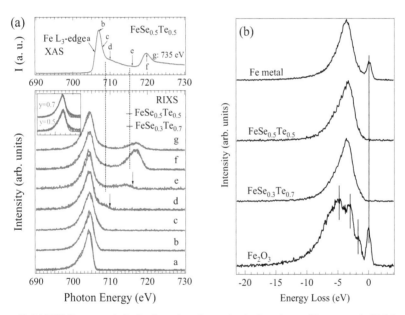

Figure 5. (a) RIXS (lower part) obtained at selected energies (a-g), as denoted by arrows in XAS (upper part). Inset shows the fluorescent feature of both samples, which overlaps well with the line at 704 eV in the non-resonant spectrum. (b) Comparison of RIXS resonantly excited at 708 eV for samples with $y=0.5$ and 0.7 and reference samples Fe_2O_3 and pure iron metal.

Figure 5(b) compares RIXS obtained with resonant excitation at 708 eV from samples with $y=0.5$ and 0.7 and from reference samples Fe_2O_3 and pure iron metal. They are displayed with an energy loss scale; elastic scattering produces the feature at an energy loss of zero. Notably, Fe_2O_3 RIXS exhibit more complex energy-loss features, which are marked by short vertical lines. These features are enhanced at particular excitation energies because of inter-electronic transitions, and are identified as *d-d* excitations (Duda et al., 2000). Fe metal yields one main fluorescent line and a more symmetric profile. Spectra of $FeSe_{1-y}Te_y$ are also obviously dominated by only a single line and resemble that of Fe metal, implying metallic character. Analysis results of the Coulombic interaction and Fe $3d$ bandwidth indicate the lack of excitation-induced energy-loss features in RIXS of iron-pnictide samples, implying a weak correlation in the iron-pnictide system (Yang et al., 2009). Therefore, known metallic iron and insulating iron oxide must be compared further to reveal the importance of electron correlation and metallicity.

Just as XAS demonstrates the density of unoccupied states (DOS) in d orbitals, non-resonant XES are interpreted simply as revealing the occupied DOS of d states, which is dominated by Fe $3d$ orbitals near the Fermi level. Figure 6 (a) shows the non-resonant Fe $L_{2,3}$-XES of Fe_2O_3, $FeSe_{1-y}Te_y$ and Fe metal. The spectra of $FeSe_{1-y}Te_y$ and Fe are nearly identical in both shape and energy, indicating that the experimental determination of the contribution of occupied Fe $3d$ states in $FeSe_{1-y}Te_y$ closely resembles that of Fe metal. The ratio of intensities of L_2 and L_3 XES should equal 1/2 for free atoms, reflecting the statistical populations of the $2p_{1/2}$ and $2p_{3/2}$ energy levels. However, in metals, Coster-Kronig transitions markedly reduce this ratio, revealing the correlations and metallicity of a system (Raghu et al., 2008). Figure 3(b) presents ratio I_2/I_3, i.e. the ratio of integrated intensities of the XES L_2 and L_3 lines. $FeSe_{1-y}Te_y$ samples with $y=0.5$ and $y=0.7$ have almost identical I_2/I_3 intensity ratios. These ratios are closer to that of correlated Fe_2O_3 than to that of the metal, indicating that the character of $FeSe_{1-y}Te_y$ more closely resembles that of correlated Fe_2O_3. This finding corresponds to that of the Fe $3d$ states in $FeSe_{1-y}Te_y$ systems, which exhibit more localized than itinerant character, as denoted by a smaller FWMH in XAS than that of Fe-pnictides.

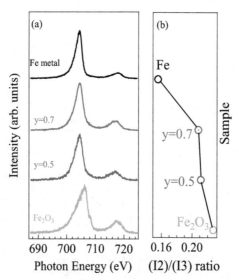

Figure 6. (a) Comparison of non-resonant Fe $L_{2,3}$-XES of Fe_2O_3, $FeSe_{1-y}Te_y$ and Fe metal. (b) Ratio of integrated intensities (I_2/I_3) of XES L_2 and L_3 lines.

The similarity between XAS and XES of $FeSe_{1-y}Te_y$ and those of Fe metal signifies the importance of metallicity in this system. This metallicity is further and directly ascertained from absorption-emission spectra, which, with a carefully calibrated energy scale, can reveal both the occupied and the unoccupied electronic densities of states around the Fermi level. Figure 7(a) displays the absorption-emission spectra, revealing the DOS of Fe $3d$ states across the Fermi level. An arrow indicates the intersection of XAS and XES, indicating that $FeSe_{1-y}Te_y$ can be regarded as metallic in nature. Figure 7(b) shows the overlain spectra of $FeSe_{1-y}Te_y$ and Fe. The Fe spectrum has a single line, which is narrower than the corresponding line of $FeSe_{1-y}Te_y$. This dominant line in the spectrum of $FeSe_{1-y}Te_y$ is slightly broader than that of Fe, perhaps owing to Se hybridization state in the lower-energy region (Yamasaki et al., 2010; Subede et al., 2008).

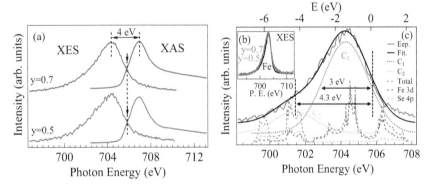

Figure 7. (a) Absorption-emission spectrum of $FeSe_{1-y}Te_y$ (y=0.5 and 0.7). Intersection yields denotes E_F. (b) Spectra of $FeSe_{1-y}Te_y$ (y=0.5 and 0.7) and Fe metal. A deviation originates from Fe-Se hybridized states. (c) Fe L_3-XES line is fitted by two components (dotted lines). The dashed lines refer to calculated DOS (from Subede et al., 2008).

Iron pnictides are weakly correlated systems, unlike high-T_c cuprates. However, a theoretical work has established that Fe-Se may not be correlated as weakly as Fe pnictides (Aichhorn et al., 2010). An important question that arises concerns whether compounds in the FeSe family exhibit a weak correlation, like that of iron pnictides, or a strong correlation, similar to that in high-T_c cuprates. Therefore, information concerning bandwidth of the Fe $3d$ states must be obtained from spectral results. An estimate of the FWHM in $FeSe_{1-y}Te_y$ XES yields a width of 4.1 eV. This value reflects the fact that core-hole lifetime and multiplet broadening, and are not associated with the Fe $3d$ bandwidth. Figure 7 (c) displays a deconvoluted spectrum that includes a single dominant line (C_1) and a low-energy contribution (C_2), which is consistent with published x-ray data (Freelon et al., 2010). The main line is associated primarily with Fe $3d$ bands; in addition, the contribution in the low-energy region is interpreted as originating from the hybridization of Fe $3d$ and Se $4p$ states (Kurmaev et al., 2009). The low-energy shoulder originates from hybridization of Fe $3d$ and As $4p$ states in the FeAsFO "1111" system. Analytical results indicate that the FeSe "11" system is simpler than the "1111" system: the low-energy contribution therefore originates from hybridization of Fe $3d$-Se $4p$ without calculation forecasts. This hybridization of Fe $3d$-Se $4p$ is also consistent with recent density functional calculations (Subede et al., 2008). According to the projected density of states (adopted from Subede et al, (2008)) in the bottom of Fig. 7(c), peak C_1 is mainly due to the Fe $3d$ states. Peak C_2 lies far below the Fermi level, and can be ascribed to the hybridized Fe $3d$-Se $4p$ states. Width of the dominant peak in Fe L_3 XES is ~3 eV. Given the instrumentation resolution, this value is taken as a guide for an upper limit of the Fe $3d$ bandwidth. Estimation results of ~3 eV for the Fe $3d$ bandwidth correlate with the on-off-resonance photoemission results, revealing that Fe dominates the binding energy range of 0-3 eV (Yamasaki et al., 2010). The low-energy contribution at 4.3 eV below E_F is attributed to Se $4p$ states, which corresponds to the on-off-resonance photoemission results (Yamasaki et al., 2010) and DOS calculation (Subede et al., 2008). The degree of electron correlation implies the relative magnitudes of the Coulomb interaction U and bandwidth W. Additionally, U is simply taken as the energy difference between the occupied and unoccupied states near E_F, and is estimated to be ~4 eV. This value matches values presented in another work, in which U of the FeSe system has been predicted to be ~4 eV (Aichhorn et al., 2010). The values imply that the magnitude of U is larger than that of the upper limit of the estimated Fe $3d$ bandwidth. The result $U/W>1$ in this work reflects a strong electronic correlation due to competition between the localized effect of U and the itinerant character of W (Zaanen et al., 1985).

Various experiments and theoretical calculations have been undertaken, especially for the "1111" and "122" systems, to elucidate electron correlation in these Fe-based compounds. Now, the results herein concerning the "11" system are compared with those obtained elsewhere for "1111" and "122" compounds. Comparing the photoemission spectrum (PES) of Fe with the Cu $2p$ core-level spectra of $CeFeAsO_{0.89}F_{0.11}$ (1111) (Bondino et al., 2008) and high-T_c cuprates ($La_{2-x}Sr_xCuO_4$ and $Nd_{2-x}Ce_xCuO_{4-\delta}$) (Steeneken et al., 2003) reveals the absence of satellite structures with higher binding energies in the Fe $2p$ spectra. This finding implies the presence of strongly itinerant Fe $3d$ electrons, which contrasts with the observation from the Cu $2p$ spectra in high-T_c cuprate. Further comparing the Fe L_3 XAS spectra reveals the lack of a well-defined multiplet structure, which is indicative of the delocalization of $3d$ band states, and subsequently providing complementary evidence of the itinerant character of Fe $3d$ electrons. Strongly itinerant $3d$ electrons are thus an important characteristic of the Fe-based superconductor, implying that it does not exhibit as strong correlations as a high-T_c cuprate. A previous study has examined the extent of electron correlations in PrFeAsO by using coupled x-ray absorption and emission spectroscopy (Freelon et al., 2010). From the spectroscopic results herein, the bandwidth of the Fe $3d$ states is estimated to be ~ 2eV, which is similar to or larger than the theoretical Coulomb parameter $U \leq 2$ eV. The relative magnitudes of U and W thus simply suggest that the "1111" system exhibits weak to intermediate electron correlations (Freelon et al., 2010). A detailed study involving the theoretical calculations for both "1111" and "122" Fe-pnictides was performed, with its results correlating well with experimental x-ray spectroscopic results (Yang et al., 2009). A cluster calculation was performed to highlight the role of strong Coulomb correlations; spectral features associated with strong correlations are shown in the theoretical results, yet are suppressed in the experimental data. Based on a direct comparison of the energy position of this feature in the experimental data and that in the cluster simulation, an upper limit of the Hubbard U of 2 eV is determined; this value is markedly smaller than the Fe $3d$ bandwidth. This finding suggests that Fe-pnictides should be viewed as a weakly correlated system. Although the XAS results herein resemble those in previous studies for Fe-pnictides and do not show the well-defined multiplet structure that is exhibited by Fe ionic oxides, two important differences are observed upon closer inspection: (a) the high-energy shoulder intensity of $FeSe_{1-y}Te_y$ is lower than that of Fe-pnictides, and (b) the linewidth of the main peak in the XAS spectrum of Fe metal is smaller than that of Fe-pnictides (Yang et al., 2009). The weak and broad shoulder is owing to hybridization of Fe $3d$-Z np states (Z: sp element). Thus, this reduction in intensity implies less hybridization than exhibited by other Fe-based "1111" and "122" systems. This finding is consistent with the observation that the linewidth of the Fe $3d$ main peak from FeSe is narrower than those from the "1111" and "122" systems. The presence of localized Fe $3d$ electrons in the FeSe system is also confirmed by XES, which can extract information about the bandwidth of the Fe $3d$ states. As is estimated, the bandwidth in this work is ~ 3 eV, although this value is larger than that, ~ 2eV, found in a previous study of the "1111" system (Bondino et al., 2008). However, according to our spectroscopic results, the Coulomb parameter U is estimated to be ~ 4eV, which exceeds values reported elsewhere, which range from 0.8 eV to 2 eV (Yang et al., 2009; Bondino et al., 2008; Nekrasov et al., 2008). Recent calculation studies also support the Coulomb parameter U~ 4eV in this work. Above differences suggest that the FeSe system differs from "1111" and "122" compounds, and the FeSe "11" system is probably not a weakly correlated system.

The electronic properties discussed above concern Fe $3d$ orbitals. The correlation between the superconducting behavior and the band structures, as well as interaction of the d-d and d-p states, may also play a vital role in these systems, as revealed by the XANES tests on $FeSe_x$ single crystals at section 3.3.1. As mentioned earlier, although doping of Te (y=0.5) enhances the Tc value, under- or over-doping decreases this value. Origin of the enhancement must be clarified. The Fe K-edge spectral profiles of the $FeSe_{1-y}Te_y$ are identical and appear to remain unaffected by Te doping (Fig. 8), which is consistent with observations of Fe L-edge XAS and RIXS. This finding suggests that the valence state and the electronic structure around the iron sites resemble each other. Additionally, the electronic configuration depends heavily on the hybridization between the orbitals of the Fe ($3d$) – Se($4p$)/Te($5p$) states. Accordingly, the electronic properties of $4p(5p)$-character, which are identified from the Se (Te) K-edge XAS spectra, must be further examined. Figures 9 (a) and 9 (b) display the Se and Te K-edge spectra, respectively. The insets highlight the energy shifts in these spectra. Se K-edge spectra have a spectral line shape resembling that in $FeSe_x$ series. The spectra are explained by a projection to the local electronic transition from the Se inner $1s$ to the outer $4p$ state. The spectra reveal two peaks, A_2 and B_2. Peak A_2 at a photon energy of around 12,658 eV originates from Fe-Se hybridized states. The peak of the doped sample has an increased intensity and shifts to a higher energy. According to related investigations (Joseph et al, 2010), the increase in peak A_2 intensity is owing to an increase in the strength of Fe $3d$-Se $4p$ hybridization. The upward shift in energy is the result of an increase in the valence states. The measured energies of the absorption edge follow the order of increasing photon energy $y=0.3$, 0.7 and 0.5, as displayed in the inset. Figure 10 plots those values. Notably, the position of the absorption edge is related to the valence state; the valence state is highest at $y=0.5$, implying that the carriers between Fe and Se are itinerant electrons in the Fe-Se hybridization band, causing a valence change upon Te substitution. This increase in the valence is an indication of the change in coordination geometry and the increase in the number of $4p$ holes. The spectra thus provide evidence that the $4p$ holes are increased the most upon Te substitution when $y=0.5$. The change in the number of holes that is caused by the variation in the strength of the Fe $3d$-Se $4p$ hybridization band is also confirmed in the x-ray absorption study of the Se deficiencies in the FeSe$_x$ system. Additionally, the number of holes changes with the superconductivity (Fig. 10) in a manner resembling that in cuprate, whose itinerant holes, via the hybridization of Cu $3d$-O $2p$ states, are responsible for superconductivity. Consequently, the change in valence implies a change in the number of holes by Fe $3d$-Se $4p$ hybridization, subsequently forming itinerant holes, as occurs in Se-deficient FeSe. Notably, the broad feature B_2 appears around 20 eV above the absorption threshold in all spectra. According to the earlier discussion of section 3.2.1 and Joseph *et al.*, (2010) this feature is attributed to multiple scatterings from symmetrical Se $4p$ states in the coordination sphere, which is related to the local structure of the Se site. This broad feature appears at the same energy in all of the FeSe$_x$ spectra and remains unaffected by the Se deficiency. This finding implies similar multiple scattering from Se $4p$ states and the same local structure around Se ions.

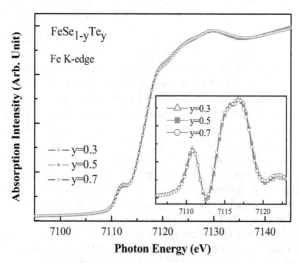

Figure 8. Fe K-edge XAS spectral profiles of FeSe$_{1-y}$Te$_y$ are identical and appear to remain unaffected by Te doping. Insets show details of 1st differential spectra, which suggest a similar electronic state around Fe site.

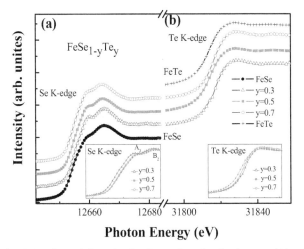

Figure 9. XAS of (a) Se K-edge and (b) Te K-edge. Insets show details of energy shifts of (a) Se K and (b) Te K-edge upon Te doping.

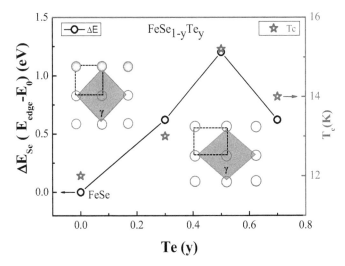

Figure 10. Energy shift of Se K-edge (black circle) and T_c as functions of Te doping. Sketch of variation of angle also shown.

However, this broad feature does not appear at the energy at which it occurred in the work of Joseph *et al.*, (2010) because of the change in the local geometry around the Se site upon Te substitution. This finding suggests local inhomogeneity, which correlates with the local inhomogeneity that was evident in XANES and EXAFS studies (Joseph et al., 2010). EXAFS analysis results indicate that the Fe-Se and Fe-Te bond lengths in $FeSe_{0.5}Te_{0.5}$ differ from each other, revealing a distinct site occupation and local in-homogeneity. A detailed polarization study of the Se K-edge demonstrates changes in the A_2 and B_2 peaks with the xy and z characteristic of the p states (Joseph et al., 2010). Owing to the natural characteristic of the p orbitals, the multiple scattering in the XAS results that are associated with p orbital symmetry reveals their different orbital orientations, thus reflecting p_{xy} and p_z character. This feature shifts to a lower energy owing to local inhomogeneity caused by doping. This finding is supported by the work of Joseph et al., (2010) and implies distortion of the tetrahedral symmetry at the Se sites. Te K-edge spectra in Fig. 9 (b) include one edge feature that is associated with 1s to 5p transitions in the coordination sphere, which reflect the local envelopment of the Te ions. The inset in Fig. 9 (b) describes details of the absorption edge. The photon energy associated with the chemical shifts increases in the order y=0.5, 0.7 and 0.3, i.e. a trend which contradicts the Se K-edge observations. The y=0.5 substitution exhibits the lowest valence state, a finding which contrasts with the Se K-edge results, possibly owing to that the charge is gained in the 5p orbital at the Te site. The critical and corresponding energy shift in the Se and Te K-edge features upon Te substitution is consequently indicative of an increase in the 4p holes and a decrease in the 5p holes at y=0.5. The tetragonal phase of FeSe has a planar sub-lattice layered structure with Se ions at the tips of the pyramid chain and an Fe plane between Se ions. The substituted Te has an ionic radius which exceeds that of Se, subsequently increasing the hybridization of Fe 3d-Se 4p/Te 5p. Comparing Se and Te K-edges reveals an expected charge transfer between Se and Te: as Te is doped into tetragonal FeSe crystals, the number of 4p holes is increased by Fe 3d–Se 4p/Te 5p hybridization. These results are consistent with an earlier study of the structural distortion that is associated with variation in the angle γ and electron-transport properties (Yeh et al., 2008). Importantly, the change in the number of p holes between Fe-Se and Fe-Te may determine superconducting behavior.

Te doping causes structural distortions in FeSe, as revealed by detailed x-ray refinement (Yeh et al., 2008). Doping expands the lattice because the ionic radius of Te exceeds that of Se. As the doping concentration increases, angle γ varies, subsequently increasing the bond length along the c-axis and altering the density of states at the Fermi level (Yeh et al., 2008), which corresponds to density-functional calculations (Subede et al., 2008). T_c and angle γ display a similar trend: both reach their maxima at y=0.5. Figure 6 plots T_c against Te doping (T_c is denoted by a red star); a simple sketch of the varying γ angle is also shown. Since the electronic structure around the Fe site in FeSe$_{1-y}$Te$_y$ does not significantly change, exactly how Se 4p holes affect superconductivity should be examined. Either the energy shift or the area under the absorption feature of XAS yields the hole concentration. Therefore, in this study, the energy shift with respect to pure FeSe is determined from the Se K-edge of FeSe$_{1-y}$Te$_y$ and is presented in Fig. 10 as a black circle. Evolution of the edge shift reasonably estimates the hole concentration. Variation in the number of Se 4p holes is closely related to the change in transition temperature. The correlation between the Se 4p hole concentration and T_c suggests that T_c depends more heavily on the variation in the number of 4p holes than on the Fe-Fe interaction in the Fe plane, a claim which is consistent with the absence of a significant change in the Fe XAS and RIXS. This finding is also consistent with the asymmetric expansion of the tetragonal lattice in relation to the structure of $FeSe_{0.5}Te_{0.5}$ – by 8 % along the c-axis, but only about 0.6% along the a-axis (Yeh et al., 2008); the Fe-Fe interaction in the Fe plane has negligible impact herein. Conversely, the role of the ligand holes and the subsequent effect of charge transfer are important. These effects may originate from Fe acting as a superexchange medium, which does not observably change from the perspective of the Fe site, a finding which is consistent with previous studies (Fang et al., 2008; Yildrim et al., 2008; Si et al., 2008).This finding may provide an important foundation for understanding the origin of superconductivity in the family of compounds considered in this study.

4. Conclusion

This study elucidates the electronic properties related to the electron correlation and superconductivity of $FeSe_x$ and $FeSe_{1-y}Te_y$, with reference to measurements of XAS and RIXS. Spectroscopic data exhibit the signature of Fe 3d localization and different hybridization effects from those of "1111" and "122" systems. The charge balance considerations from p-hole also result in itinerant electrons. Fluctuation in the number of ligand 4p holes may arise from the charge transfer between Se and Te in the $FeSe_{1-y}Te_y$ crystals. Analysis results indicate that the superconductivity in Fe-based compounds of this class is strongly associated with the ligand 4p hole state. Additionally, the variation of Tc correlates well with the structural deformation and the change in the Se 4p holes. Moreover, the symmetry of Fe in the ab plane changes from the 4p orbital to modulating (varying) coordination geometry. XRD measurements indicate that this lattice distortion that increases with Se deficiency and the Te doped. Tetragonal FeSe with a PbO structure not only has the same planar sub-lattice as layered Fe-based quaternary oxypnictides, but also exhibits a structural stability upon Te substitution; it is a promising candidate for determining the origin of Tc in Fe-based superconductors. A fundamental question concerning the role of Fe magnetism in these superconductors is yet to be answered. The importance of charge transfer and the ligand 4p hole state should be considered as well.

Author details

Chi Liang Chen and Chung-Li Dong
Institute of Physics, Academia Sinica , Nankang, Taipei,
National Synchrotron Radiation Research Center (NSRRC), Hsinchu, Taiwan

Acknowledgement

The authors would like to thank the National Science Council of the Republic of China, Taiwan (Contract Nos. NSC-98-2112-M-213-006-MY3 and NSC-099-2112-M-001-036-MY3) for financially supporting this research. M. K. Wu, Y. Y. Chen, S. M. D. Rao, and K. W. Yeh at Academia Sinica are appreciated for providing study samples and their valuable discussions. J.-F. Lee, T. S. Chan, C. W. Pao, J. M. Chen, and J. M. Lee of NSRRC are commended for their valuable discussions and experimental support. J.-H. Guo and W. L. Yang of Advanced Light Source are gratefully acknowledged for their experimental support.

5. References

Aichhorn, M. Biermann, S, Miyake, T., Georges A. & Imada, M. (2010). Theoretical evidence for strong correlations and incoherent metallic state in FeSe. *Physical Review B*, 82: 064504.

Aichhorn, M., Pourovskii, L., Vildosola, V., Ferrero, M., Parcollet, O., Miyake, T., Georges, A. & Biermann, S. (2009). Dynamical mean-field theory within an augmented plane-wave framework: Assessing electronic correlations in the iron pnictide LaFeAsO. *Physical Review B*, 80:085101.

Arčon, D., Jeglič, P., Zorko, A,. Potočnik, A., Ganin, A.Y., Takabayashi, Y., Rosseinsky, M.J. & Prassides. K. (2010). Coexistence of localized and itinerant electronic states in the multiband iron-based superconductor $FeSe_{0.42}Te_{0.58}$., *Physical Review B*, 82:140508(R)

Bondino, F., Magnano, E., Malvestuto, M., Parmigiani, F., McGuire, M.A., Sefat, A.S., Sales, B.C., Jin, R., Mandrus, D., Plummer, E.W., Singh, D.J. & Mannella, N. (2008). Evidence for strong itinerant spin fluctuations in the normal state of $CeFeAsO_{0.89}F_{0.11}$ iron-oxypnictide superconductors, *Physical Review Letters*, 101: 267001-267004.

Chang, C.L., Chen, C.L., Dong, C.L., Chern, G., Lee, J.-F., & Jang, L.Y. (2001). X-ray absorption near edge structure studies of $Fe_{1-x}Ni_xO_y$ thin films *Journal of Electron Spectroscopy and Related Phenomena*, 114-116:545-548.

Cruz, C. de la, Huang, Q., Lynn, J.W., Li, Ratcliff II, J.W., Zarestky, J.L., Mook, H.A., Chen, G.F., Luo, J.L., Wang, N.L. & Dai, P. (2008). Magnetic order close to superconductivity in the iron-based layered $LaO_{1-x}F_xFeAs$ systems. *Nature (London)*, 453:899-902.

de Groot, F., Vankó, G. & Glatzel, P., (2009). The 1s x-ray absorption pre-edge structures in transition metal oxides. *Journal of physics: Condensed Matter*, 21: 104207.

Dong, J., Zhang, H.J., Xu, G., Li, Z., Li, G., Hu, W.Z., Wu, D., Chen, G.F., Dai, X., Luo, J.L., Fang, Z. & Wang, N.L. (2008). Competing orders and spin-density-wave instability in $La(O_{1-x}F_x)FeAs$. *Europhysics Letters*, 83:27006.

Duda, L.C., Nordgren, J., Dräger, G., Bocharov, S. & Kirchner, T. (2000). Minimal two-band model of the superconducting iron oxypnictides, *Journal of Electron Spectrosocpy and Related Phenomena*, 110: 275-285.

Fang, M.H., Pham, H.M., Qian, B., Liu, T.J., Vehstedt, E.K., Spinu, L. & Mao, Z.Q. (2008). Superconductivity close to magnetic instability in $Fe(Se_{1-x}Te_x)_{0.82}$, *Physical Review B*, 78: 224503-224507.

Freelon, B., Liu. Y.S., Rotundu, C.R., Wilson, S.D., Guo, J, Chen, J.L., Yang, W.L., Chang, C.L., Glans.P.A., Shirage, P., Iyo, A.& Brigeneau, R.J. (2010). X-ray absorption and emission spectroscopy study of the effect of doping on the low energy electronic structure of $PrFeAsO_{1-\delta}$ *Journal of the Physical Society of Japan*, 79: 074716-1-074716-7.

Grioni, M., Goedkoop, J.B., Schoorl, R., de Groot, F.M.F., Fuggle, J.C., Schäfers, F., Koch, E.E., Rossi, G., Esteva, J.-M. & Karnatak, R.C. (1989). Studies of copper valence states with Cu L_3 x-ray-absorption spectroscopy, *Physical Review B*, 39: 1541-1545.

Hsu, F.C., Luo, J.Y., Yeh, K.W., Chen, T. K., Huang, T.W., Wu, P.M., Lee, Y.C., Huang, Y. L., Chu, Y.Y., Yan, D.C. & Wu, M.K. (2008). Superconductivity in the PbO-type structure α-FeSe. *Proceedings of the National Academy of Sciences (PNAS)*, 105:14262-14264.

Joseph, B., Iadecola, A., Puri, A., Simonelli, L., Mizuguchi, Y., Takano, Y. & Saini, N.L. (2010). Evidence of local structural inhomogeneity in $FeSe_{1-x}Te_x$ from extend x-ray absorption fine structure, *Physical Review B*, 82: 020502- 020505(R).

Joseph, B., Iadecola, A., Simonelli, L., Mizuguchi, Y., Takano, Y., T. Mizokawa, T. & Saini, N.L. (2010). A study of the electronic structure of $FeSe_{1-x}Te_x$ chalcogenides by Fe and Se K-edge x-ray absorption near edge structure measurements, *Journal of Physics: Condensed Mater*, 22: 485702485706.

Kamihara, Y., Watanabe, T., Hirano, M. & Hosono, H. (2008). Iron-Based Layered Superconductor $La[O_{1-x}F_x]FeAs$ (x=0.05–0.12) with T_c = 26 K. *Journal of The American Chemical Society*, 130:3296-3297.

Kroll, T., Bonhommeau, S., Kachel, T., Dürr, H.A., Werner, J., Behr, G., Koitzsch, A., Hübel, R., Leger, S., Schönfelder, R., Ariffin, A.K., Manzke, R., de Groot, F.M.F., Fink, J., Eschrig, H., Büchner, B. & Knupfer, M. (2008). Electronic structure of $LaFeAsO_{1-x}F_x$ from x-ray absorption spectroscopy. *Physical Review B*, 78:220502.

Kurmaev, E.V., McLeod, J.A., Buling, A., Skorikov, N.A., Moewes, A., Neumann, M., Korotin, M.A., Izyumov, Y.A., Ni, N. & Candield. P.C. (2009). Contribution of Fe 3d states to the Fermi level of $CaFe_2As_2$, *Physical Review B* Vol. 80: 054508-054513.

Lee, K.-W., Pardo, V. & Pickett, W.E. (2008). Magnetism driven by anion vacancies in superconducting $α-FeSe_{1-x}$, *Physical Review B*, 78: 174502-174506.

Longa, S.D., Arcovito, A., Vallone, B., Castellano, A.C., Kahn, R., Vicat, J., Soldo, Y. & Hazemann, J.L. (1999). Polarized X-ray absorption spectroscopy of the low-temperature photoproduct of carbonmonoxy-myoglobin, *Journal of Synchrotron Radiation*, 6: 1138-1147.

Lytle, F. W., Greegor, R.B., Sandstrom, D.R., Marques, E.C., Wong, J., Spiro, C.L., Huffman, G.P. & Huggins, F. E. (1984). Measurement of soft x-ray absorption spectra with a fluorescent ion chamber detector. *Nuclear Instruments and Methods in Physics Research Section A: Accelerators, Spectrometers, Detectors and Associated Equipment*, 226:542-548.

Ma, F., Ji, W., Hu, J., Lu, Z.Y. & Xiang, T. (2009). First-Principles Calculations of the Electronic Structure of Tetragonal α-FeTe and α-FeSe Crystals: Evidence for a Bicollinear Antiferromagnetic Order. *Physical Review Letters*, 102:177003

Malaeb, W., Yoshida, T., Kataoka, T., Atsushi, Fujimori, A., Kubota, M., Ono, K., Usui, H., Kuroki, K., Arita, R. & Hosono, H. (2008). Electronic Structure and Electron Correlation in $LaFeAsO_{1-x}F_x$ and $LaFePO_{1-x}F_x$. *Journal of the Physical Society of Japan*, 77:093714.

McQueen, T.M., Huang, Q., Ksenofontov, V., Felser, C., Xu, Q., Zandbergen, H., Hor, Y.S., Allred, J., Williams, A.J., Qu, D., Checkelsky, J., Ong,. N.P. & Cava, R.J. (2009). Extreme sensitivity of superconductivity to stoichiometry in $Fe_{1+\delta}Se$. *Physical Review B*, 79:014522.

Merz, M., Nücker, N., Schweiss, P., Schuppler, S., Chen, C.T., Chakarian, V., Freeland, J., Idzerda, Y.U., Kläser M., Müller-Vogt, G. & Wolf, Th. (1998). Site-specific x-ray absorption spectroscopy of $Y_{1-x}Ca_xBa_2Cu_3O_{7-y}$: Overdoping and role of apical oxygen for high temperature superconductivity, *Physical Review Letters*, 80: 5192-5195.

Mizuguchi, Y., Tomioka, F., Tsuda, S., Yamaguchi, T. & Takano, Y. (2008). Superconductivity at 27 K in tetragonal FeSe under high pressure, *Applied Physics Letter*, 93: 152505-152507.

Mok, B.H., Rao, S.M., Ling, M.C, Wang, K.J., Ke, C.T., Wu, P.M., Chen, C.L., Hsu, F.C., Huang, T.W., Luo, J.Y., Yan, D.C., Yeh, K.W., Wu, T.B., Chang, A.M. & Wu, M.K. (2009). Growth and Investigation of Crystals of the New Superconductor α-FeSe from KCl Solutions. *Crystal Growth and Design*, 9(7): 3260-3264.

Nakayama, K., Sato, T., Richard, P., Kawahara, T., Sekiba, Y., Qian, T., Chen, G.F., Luo, J.L., Wang, N.L., Ding, H.& Takahashi, T. (2010). Angle-Resolved Photoemission Spectroscopy of the Iron-Chalcogenide Superconductor $Fe_{1.03}Te_{0.7}Se_{0.3}$: Strong Coupling Behavior and the Universality of Interband Scattering. *Physical Review Letters*, 105:197001.

Nekrasov, I.A., Pchelkina, Z.V. & Sadovskii, M.V. (2008). Electronic structure of prototype AFe_2As_2 and ReOFeAs high-temperature superconductors: A comparison, *JETP Letters*, 88: 144-149.

Norgdren, J., Bray, G., Cramm, S., Nyholm, R., Rubensson, J.E. & Wassdahl, N. (1989). Soft x-ray emission spectroscopy using monochromatized synchrotron radiation (invited). *Review of Scientific Instruments*. 60:1690.

Pourret, A., Malone, L. Antunes, A. B., Yadav, C.S., Paulose, P.L., Fauque, B. & Behnia, K. (2010). Strong correlation and low carrier density in $Fe_{1+y}Te_{0.6}Se_{0.4}$ as seen from its thermoelectric response. *Physical Review B*, 83:020504.

Raghu, S., Qi, X.L., Liu, C.X., Scalapino, D.J. & Zhang, S.C. (2008). Minimal two-band model of the superconducting iron oxypnictides, *Physical Review B*, 77: 220503-220506 (R).

Ren, Z.-A., Lu, W., Yang, J., Yi W., Shen X.-L., Li Z.-C., Che G.-C., Dong X.-L., Sun L.-L., and Zhou F., Zhao Z.-X (2008), Superconductivity at 55 K in iron-based F-doped layered quaternary compound $Sm[O_{1-x}F_x]FeAs$. *Chinese Physics Letters*, 25:2215-2216.

Sales, B.C., Sefat, A.S., McGuire, M.A., Jin, R.Y., Mandrus, D. & Mozharivskyj, Y. (2009). Bulk superconductivity at 14 K in single crystals of $Fe_{1+y}Te_xSe_{1-x}$. *Physical Review B*, 79, 094521.

Si, Q. & Abrahams, E. (2008). Strong correlations and magnetic frustration in the high T_c iron pnictides, *Physical Review Letters*, 101: 076401-076404.

Steeneken, P.G., Tjeng, L.H., Sawatzky, G.A., Tanaka, A., Tjernberg, O., Ghiringhelli, G., Brookes, N.N., Nugroho, A.A. & Menovsky A.A. (2003).Crossing the gap from p- to n-type doping: nature of the states near the chemical potential in $La_{2-x}Sr_xCuO_4$ and $Nd_{2-x}Ce_xCuO_{4-\delta}$, *Physical Review Letters* Vol. 90: 247005-247008.

Subedi, A., Zhang, L., Singh, D.J. & Du, M.H. (2008). Density functional study of FeS, FeSe, and FeTe: Electronic structure, magnetism, phonons, and superconductivity, *Physical Review B* Vol. 78: 134514-134519.

Takahashi, H., Igawa, K., Arii, K., Kamihara, Y., Hirano, M. & Hosono, H. (2008). Superconductivity at 43 K in an iron-based layered compound $LaO_{1-x}F_xFeAs$. *Nature (London, United Kingdom)*, 453:376-378.

Tamai, A., Ganin, A.Y., Rozbicki, Bacsa, E.J., Meevasana, W., King, P.D.C., Caffio, M., Schaub, R., Margadonna, S., Prassides, K., Rosseinsky, M.J. & Baumberger, F. (2010). Strong Electron Correlations in the Normal State of the Iron-Based $FeSe_{0.42}Te_{0.58}$ Superconductor Observed by Angle-Resolved Photoemission Spectroscopy. *Physical Review Letters*, 104:097002

Wu, M.K., Hsu, F.C., Yeh, K.W., Huang, T.W., Luo, J.Y., Wang, M.J., Chang, H.H., Chen, T.K., Rao, S.M., Mok, B.H., Chen, C.L., Huang, Y.L., Ke, C.T., Wu, P.M., Chang, A.M., Wu, C.T. & Perng, T.P. (2009). The development of the superconducting PbO-type β-FeSe and related compounds. *Physica C*, 469 (9-12):340-349.

Xu, G., Ming, W.M., Yao, Y.G., Dai, X., Zhang, S.C. & Fang, Z. (2008). Doping-dependent phase diagram of LaOMAs (M=V–Cu) and electron-type superconductivity near ferromagnetic instability. *Europhysics Letters*, 82:67002.

Yamasaki, A., Matsui, Y., Imada, S., Takase, K., Azuma, H., Muro, T., Kato, Y., Higashiya, A., Sekiyama, A., Suga, S., Yabashi, M., Tamasaku, K., Ishikawa, T., Terashima, K., Kobori, H., Sugimura, A., Umeyama, N., Sato, H., Hara, Y., Miyagawa, N. & Ikeda I. (2010). Electron correlation in the FeSe superconductor studied by bulk-sensitive photoemission spectroscopy. *Physical Review B*, 82: 184511.

Yang, W.L., Sorini, A.P., Chen, C.-C., Moritz, B., Lee, W.-S., Vernay, F., Olalde-Velasco, P., Denlinger, J.D., Delley, B., Chu, J.-H., Analytis, J.G., Fisher, I.R., Ren, Z.A., Yang, J., Lu, W., Zhao, Z.X., van den Brink J., Hussain, Z., Shen, Z.-X. & Devereaux, T.P. (2009). Evidence for weak electronic correlations in iron pnictides. *Physical Review B*, 80:014508.

Yeh, K.W., Huang, Z.W., Huang, Y. L., Chen, T. K., Hsu, F. C., Wu, P.M., Lee, Y. C., Chu, Y. Y., Chen, C. L., Luo, J. Y., Yan,D.C. & Wu, M.K. (2008). Tellurium substitution effect on superconductivity of the α-phase iron selenide. *Europhysics Letters*, 84:37002.

Yeh, K.W., Ke, C.T., Huang, T.W., Chen, T.K., Huang, Y.L., Wu, P.M. & Wu, M.K. (2009). Superconducting $FeSe_{1-x}Te_x$ Single Crystals Grown by Optical Zone-Melting Technique. *Crystal Growth Design*, 9:4847.

Yildrim, T. (2008). Origin of the 150-K anomaly in LaFeAsO: competing antiferromagnetic interactions, frustration, and a structural phase transition, *Physical Review Letters*, 101: 057010-057013.

Yoshida, R., Wakita, T., Okazaki, H., Mizuguchi, Y., Tsuda, S., Takano, Y., Takeya, H., Hirata, K., Muro, T., Okawa, M., Ishizaka, K., Shin S., Harima, H., Hirai M., Muraoka, Y.,

& Yokoya, T. (2009). Electronic Structure of Superconducting FeSe Studied by High-Resolution Photoemission Spectroscopy. *The Physical Society of Japan*, 78: 034708.

Zaanen, J., Sawatzky, G.A. & Allen, J.W. (1985). Band gaps and electronic structure of transition-metal compounds, *Physical Review Letters*, 55: 418-421.

Chapter 4

Improvement of Critical Current Density and Flux Trapping in Bulk High-T$_c$ Superconductors

Mitsuru Izumi and Jacques Noudem

Additional information is available at the end of the chapter

1. Introduction

The present chapter describes an overview of flux trapping with enhancement of the critical current density (J_c) of a melt-growth large domain (RE)Ba$_2$Cu$_3$O$_{7-d}$, where RE is a light rare earth ions such as Y, Gd or Sm. These high-T$_c$ superconductor bulks have attracted much interest for a variety of magnet applications, since high density and large volume materials potentially provide an intensified magnetic flux trapping, thanks to the optimized distribution of pinning centres. The melt growth process and material processing to introduce well-defined flux pinning properties are overviewed. As a first step, we summarize an effort to achieve a growth of homogeneous large grains with the second phase RE211 in the RE123/Ag matrix. RE-Ba-Cu-O material has a short coherence length and a large anisotropy, and thus any high-angle grain boundary acts as a weak link and seriously reduces the critical current density [1, 2]. In engineering applications, high texture and c-axis-orientated single grains/domains are required. Large-sized, high-performance RE-Ba-Cu-O single grains are now commercially available. The trapped flux density (B_{trap}) due to flux pinning or associated superconducting currents flowing persistently in a RE-Ba-Cu-O grain is expressed in a simple model, such as:

$$B_{trap} = A\mu_0 J_c r,$$

where A is a geometrical constant, μ_0 is the permeability of the vacuum and r is the radius of the grain [1]. There are two approaches to enhancing the trapped flux of the grain. One is to enhance the critical current density and the other is to increase the radial dimension of the crystals. Increasing the dimension requires the formation of homogeneous grain growth, and the enhancement of the critical current density is encouraged with the improvement of flux pinning properties.

The top-seeded melt-growth (TSMG) method has been widely used to fabricate large, single-grain RE-Ba-Cu-O superconducting bulks that show a considerable ability in magnetic flux trapping and great potential for large-scale applications [1]. Hot seeding and cold seeding procedures have been studied. For hot-seeding processes, Nd-Ba-Cu-O or Sm-Ba-Cu-O crystals with a high decomposition temperature are put on the matrix during the growth of the bulk, around a peritectic temperature (T_P), which is not convenient for the batch process and often brings problems for reproducibility. Cardwell et al. have introduced a cold-seeding process with Mg-doped Nd-Ba-Cu-O crystals as generic seeds whose decomposition temperature is higher than the pure substance [3, 4]. Nd-Ba-Cu-O and Sm-Ba-Cu-O thin films grown on MgO substrates have been examined as cold seeds [5, 6]. Thanks to the superheating phenomenon of Nd-Ba-Cu-O thin films, the maximum temperature (T_{max}) is increased up to even 1090 °C [7, 8]. Muralidhar et al. have reported a batch process of Gd-Ba-Cu-O bulks [9, 10]. Recently, it has been reported that a buffer pellet inserted between the seed and the matrix effectively suppresses the chemical contamination caused by the dissolution of the seed, without affecting the texture growth, and the T_{max} is increased to 1096 °C [8, 11].

An idea for the novel cold-seeding of a top-seeded melt-growth with a RE-Ba-Cu-O bulk has been worked on by employing an MgO crystal seed and a buffer pellet [12]. The growth process is composed of two stages. The MgO seed was for the texture-growth of the small RE-Ba-Cu-O pellet with a high melting point (T_P), and the textured pellet induced the texture growth of the bulk at a lower temperature. Undercooling and the RE211 content of the pellet were adjusted to avoid the misorientation caused by lattice mismatch between MgO and the RE-Ba-Cu-O matrix. Bulk samples prepared with this method show good growth sections and superconducting performance. One of the promising advantages of this method is in the processing of high T_P RE-Ba-Cu-O bulks with a cold seeding method, for example Nd-Ba-Cu-O bulks.

Detailed information for the preparation of the samples is described elsewhere [12]. GdBa$_2$Cu$_3$O$_{7-\delta}$ (3N, Gd123), Gd$_2$BaCuO$_5$ (3N, Gd211) powders were employed with 40 mol% of Gd211 for Gd123. 10 wt.% Ag$_2$O and 0.5 wt.% Pt were added. A small buffer pellet of Gd123 contained a certain amount of Gd211. A single (100)-oriented MgO seed was placed onto the small pellet.

According to the results of the differential thermal analysis (DTA) measurements [12], we used the heat treatment profile shown in Fig. 1. The sample was heated within 10 hours to T_{max}, 90 °C higher than the $T_{p\text{-matrix}}$ (for the matrix with the addition of Ag$_2$O). After one hour, the temperature was reduced to $T_{p\text{-buffer}} - \Delta T$ within 30 minutes so as to begin the growth of the buffer pellet. ΔT stands for the undercooling. After that the temperature was reduced over 30 minutes to $T_{p\text{-matrix}}$ and further slowly decreased by 30 °C with a cooling rate of 0.3 °C/h. Eventually, the temperature was decreased to room temperature within 10 hours. The following post-annealing process has been reported in our previous studies [13, 14].

Fig. 2 shows the appearance of the bulk samples prepared by conventional hot-seeding (a), cold-seeding using a Nd123 thin film (b), and cold-seeding in association with a MgO-buffer pellet (c). The c-axis oriented single-grain growth for the buffer pellet is of importance. Cardwell et al. and Babu et al. have reported that the geometry of Nd-Ba-Cu-O single grains, texture-processed by MgO seeds, will vary from rectangular ($\Delta T < 10$ °C) to rhombohedral (high values of ΔT) under different growth temperatures [15, 16] Cima et al. have demonstrated that the so called "faced plane growth front" type of solidification interface morphologies is largely dependent on the growth rate [17]. The undercooling is directly related to the growth rate. Meanwhile, the RE211 content affects the growth rate. A slow cooling between 1030 °C and 1025 °C with 10 mol % Gd211 content is suitable for the buffer pellet's texture growth.

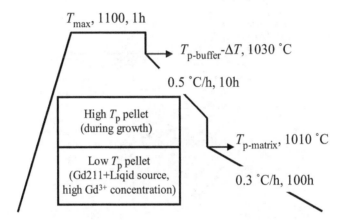

Figure 1. Schematic illustration of thermal profile for the cold-seeding growth of Gd-Ba-Cu-O bulk superconductors [12].

The present cold-seeding method can be used for growth with a high T_P bulk in the air, for example in a Nd-Ba-Cu-O bulk. The same progress for detecting a suitable undercooling and Nd422 content has been carried out in the Nd-Ba-Cu-O system. A slow cooling between 1067 °C and 1062 °C with rate of 0.5 °C/h and 10 mol% Nd422 content have been proven to offer the best growth conditions for the Nd-Ba-Cu-O buffer pellet [12].

The growth can be transferred from a high-T_P pellet to a low-T_P pellet. As illustrated in Fig. 1, during the growth of the high-T_P part, the low-T_P part is kept at a relatively high temperature, which means that a high RE concentration may exist. It is promising for extending the growth window and benefits of a larger scale bulk superconductor from the viewpoint of homogeneity. The addition of silver as well as the mixture of two or three kinds of RE123 powders may change T_P. Recently, we have found that by doping 30 mol% Nd123 into the Gd123 precursor powders, the T_P is increased by 6 °C while keeping texture growth.

Fig. 3 shows the microstructure of the portion at the buffer/matrix interface. The boundary is denoted by the broken line. A different contrast of Gd211 density was observed below and above the boundary. Because of the push effect of Gd211, a high Gd211 density area is formed at the interface. The composition of the matrix was measured by EPMA for points indicated by green closed circles in Fig. 3 (a). There is $Gd_{1+0.02}Ba_{2-0.02}Cu_{2.72}$ in the composition of the buffer side and it is approximately close to $Gd_{1+0.09}Ba_{2-0.09}Cu_{2.64}$ in the matrix side. We suspect that because of the large undercooling and growth rate, the Gd/Ba substitution is inhibited at the buffer side.

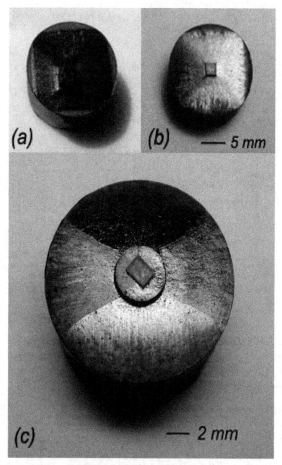

Figure 2. Gd-Ba-Cu-O single-grain bulk of 16 mm in diameter prepared by (a) a hot-seeding process, (b) a cold-seeding process using a Nd123 thin film seed, (c) a cold-seeding growth with a MgO crystal seed and a buffer pellet [12].

Secondly, we emphasize how to launch additional pinning centres into the RE123/Ag matrix. There are several strategies which are partly analogue to the implantation of pinning centres in thin film forms. Partial atomic substitutions of the Ba^{2+} site with RE^{3+} in RE123 induce a so-called "peak effect" around 1.5-2.0 T in the J_c-B curves. The substitution of 1D Cu site in the RE123 structure with other ions results in an enhanced peak effect [18]. Many kinds of additions of non-superconducting metal oxides have been studied in the Gd123 /Ag

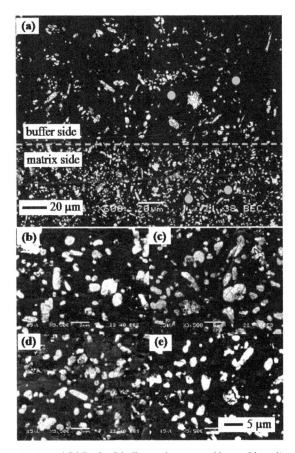

Figure 3. The microstructure of Gd-Ba-Cu-O bulk sample processed by a cold-seeding method using a MgO-buffer pellet observed by SEM. (a) Buffer/matrix interface, (b) C1: under the seed, (c) B1: periphery in the growth sector, (d) C4 position [12], (e) B4 position [12].

matrix with Gd211. Gd211 tends to form domains of a large size inside Gd123. Various kinds of oxides and RE$_2$Ba$_4$MCuO$_{11}$ (RE2411 particles, M = Zr, U, Mo, W, Ta, Hf, Nb) are introduced into the RE-Ba-Cu-O matrix as second phase particles so as to enhance flux pinning [19-20]. Up to now, the record of J_c reaches 640 kA/cm^2 and 400 kA/cm^2 at 77 K in the self field and 2 T, respectively. This record was achieved in the (Nd,Eu,Gd)-Ba-Cu-O bulk combining the benefits of dense regular arrays of a RE-rich RE123 solid solution, the initial Gd211 particles that were 70 nm in diameter and the formed small (< 10 nm) Nb (or Mo, Ti)-based nanoparticles [21]. Systematic research of the doping effect has been also carried out in our laboratory. J_c of 100 kA/cm^2, 68 kA/cm^2 and 80 kA/cm^2 were obtained at 77 K in a self field by doping with ZrO$_2$, ZnO and SnO$_2$ particles, respectively [22, 23]. It is interesting that the addition of nano-sized metal oxides - such as SnO and/or ZrO$_2$, for example - provides not only the simple *in situ* formation of $BaSnO_3$ and $BaZrO_3$ but also the fining of the size of Gd211 distributed inside the matrix, as shown in Fig. 4. These effects are classified with the *in situ* formation of the nano-sized flux pinning centres during the growth process. To make for strong flux pinning, the introduction of nano-sized inclusions in textured bulk HTSs constitutes an effective means. Apart from making a fine second phase particle, dilute impurity doping is even more important for improving flux pinning. The increased Jc in such a dilute doping bulk is even several times larger than that in the reference sample. Therefore, at present, we focus on the chemical approach of dilute impurity doping. Different additives such as BaO_2, ZrO_2, ZnO, NiO, SnO_2, Co_3O_4, Fe_3O_4, Ga_2O_3 and Fe-B alloy [20, 22,23, 24-30], which lead to a slight decrease of T_c in the bulk RE-123, except for a few kinds of additives like Gd2411 and titanium oxide.

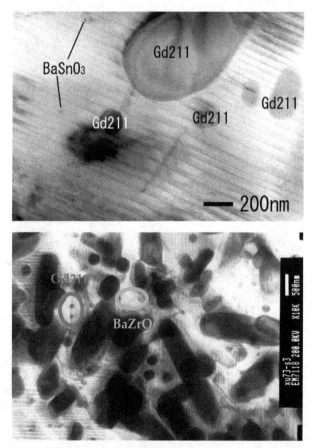

Figure 4. The microstructure of Gd-Ba-Cu-O bulk sample processed by the hot-seeding method observed by SEM. (a) ZrO_2 addition [22] and (b) SnO_2 addition [23].

Cardwell et al. [4] and Muralidhar et al. [10] have developed general process routes to grow batches of RE-Ba-Cu-O single domain superconductors with good pinning performance. The flux pinning and J_c performance of a RE-Ba-Cu-O bulk yield remarkable improvements by dispersing the non-superconducting secondary pinning phase into the RE-123 matrix. Successful attempts have been made to add nano-sized impurities [22, 23, 31], the fined RE-211s [32, 33] and Pt, Ce additives to prevent the Ostwald ripening of RE-211 inclusions into the precursors and to enhance flux pinning. On the other hand, compared with core pinning by normal non-superconducting particles, the use of ferromagnetic pinning centres results in interactions between a magnetic dipole moment and flux lines, which yields a potential Upin proportional to $-mb$, where m is the moment of magnetic dipole and b is the field of the vortex at the distance of the dipole [34]. The deeper potential wells may reduce the Lorentz force on the vortices [35-37].

Xu Yan et al. have found that Fe-B quenched amorphous magnetic alloy particles with small amounts of Cu-Nb-Si-Cr may be a useful additive for flux trapping properties [25, 26]. The results show that the J_c was enhanced under both low- and high-magnetic fields with the addition of 0.4 mol% of Fe-B particles [25, 26, 38].

SEM observations were also carried out to confirm the information of the Fe-rich region obtained from TEM. The representative back scattered electron image is shown in Fig. 5 (a), where the larger particles represent silver, and the homogeneous distributed small particles are Gd-211 embedded in the Gd-123 matrix, according to the EDX analysis. Consistent with the results from TEM, the Fe element was only found in the vicinity of silver, as shown in Fig. 5(b). This may be attributed to the following three reasons:

First, silver and Fe_3O_4 possess a cubic structure with a lattice mismatch: $a = 8.397$ A for Fe_3O_4 and $a = 4.090$ A for silver. Two unit cells of silver may be nearly equi-length with that of one unit cell of Fe_3O_4, giving a small lattice mismatch of 2.65%. Second, the oxidization temperature of Fe-B additives obtained in our DTA results is identified at around 960 °C, very close to the melting point of silver 961.9 °C. Meanwhile, the Fe-B additives were oxidized into Fe-containing components with porous structures, as confirmed by our experiment of annealing Fe-B alloy separately. As a result, the melted silver may fill these voids at high temperatures. Third, this oxidation process is an exothermal reaction, which accelerates the melting of adjacent silver particles. Fourth, the released oxygen from Ag_2O would be the source of the oxidization of Fe-B additives. The release of oxygen might provide a potential channel for the flowing of melted silver to Fe_3O_4. The advantage of the present materials process is in eliminating the proximity effect between magnetic Fe_3O_4 and the Gd-123 superconducting matrix by the silver as a buffer layer.

Besides this, in the growth process, the added Pt may exist around the boundary between Gd123 and Ag, for example. Fe is known to be with Ag. The addition of magnetic oxide, such as Fe_2O_3 or other kinds of Fe alloys, has been investigated from the viewpoint of the magnetic pinning effect. Tsuzuki et al. have reported that Fe_2O_3 was introduced into the Gd123 matrix [39]. The maximum trapped flux increased by over 30 %. In the case of Fe-B particles addition, J_c is increased in both center and edge of the samples. However, no enhancement of J_c was observed at the edge with the Fe_2O_3 addition. Here, there is the difference of the integrated flux between Fe-B addition and Fe_2O_3 addition from the spatial distribution of J_c. The origin of homogeneous J_c and the effect of in-situ formation of Fe_2O_3 in the Fe-B added Gd123 bulks are the keys to improve the performance of the magnetic field trapping [40].

Separately, the optimal addition of these magnetic particles induces an increase of the number of Gd211 particles while decreasing the size. We emphasize the current issues concerning the homogeneity of the distribution of these particles together with TEM observations [38].

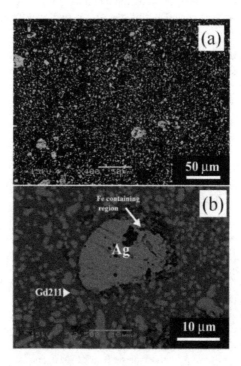

Figure 5. (a) Low magnification of an SEM image of a 1.4 mol% Fe-B doped Gd-Ba-Cu-O C1 specimen. (b) SEM image of a Fe-rich region [26].

Another unique aspect concerning flux trapping is to distribute holes drilled within the bulk pack. The recently reported [41-45] hole-patterned $YBa_2Cu_3O_y$ (Y123) bulks with improved superconducting properties are highly interesting from the points of view of material quality and their variety of application. It is well known that the core of plain bulk superconductors needs to be fully oxygenated, and some defects like cracks, pores and voids [46, 47] must be suppressed in order that the material can trap a high magnetic field or else carry a high current density. Some previous studies [48-51] demonstrated that, by filling the cracks, enhancing thermal conductivity or by reinforcing the YBCO bulk material, the properties can be improved and a trap field of up to 17 T at 29 K can be reached. One of the interests of this new sample geometry is in increasing specific areas for thermal exchange, shortening the oxygen diffusion path, and offering the possibility of reinforcing the superconductor materials. To minimize the above defects, we propose the improvement of the superconducting material with an innovative approach - "material by design" based on the concept of a YBa$_2$Cu$_3$O$_y$ (Y123) bulk with multiple holes.

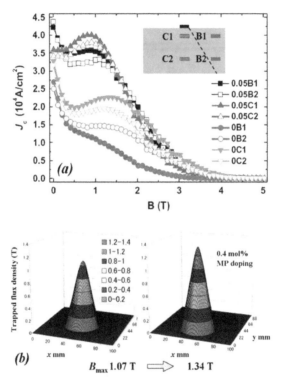

Figure 6. (a) The J_c-B curves of specimens cut from different positions of MP doped and un-doped bulk samples. (b) The trapped magnetic field of 46-mm MP-doped and un-doped bulks [25].

The details of the multiple holes process of YBa$_2$Cu$_3$O$_y$ (Y123) are reported elsewhere [41, 42]. Basically, the holes in the pre-sintered bulk were realized by drilling cylindrical cavities with different diameters, 0.5-2 mm through the circular or square shaped sample. The holes are arranged in a regular network on the plane of the samples. A SmBa$_2$Cu$_3$O$_x$ (Sm123) seed used as a nucleation centre was placed (between the holes close to the centre) on the top so as to obtain the single domain of the samples. The seed orientation was chosen to induce a growth with the c-axis parallel to the pellet axis. The elaboration of single domains through the drilled pellets is then conducted in a manner similar to the plain pellets. But how to claim a single domain on the drilled sample? The demonstration of the growth of single domains from the perforated structure is shown by Fig. 7(a). The growth lines of faceted growth on the surface of the drilled single domain half are not clearly observed, but they exist when compared to the plain half. This shows that the pre-formed holes do not seem to

(a)

(b)

Figure 7. (a). Macrograph of the single domain pellet sample where only half has been drilled [44]. (b). Pictures of the surfaces of a drilled (left) and a plain (right) pellet taken at an intermediate stage of the growth process. The bright square is the growing domain with a seed at its centre. The steps and streaks result from the interaction of the holes with the growth front (left) [44].

disturb the growth of the single domain, which is confirmed by the top seed melt growth process of other perforated samples prepared by Chaud et al. [42]. Basically, the ability of a growth front to proceed through an array of holes or a complex geometry is not evident *a priori*. *In situ* video monitoring of the surface growth confirms that it proceeds as for a plain pellet. The growth starts from the seed. A square pattern typical of the growth front of the tetragonal Y123 phase in the a- and b- directions appears below the seed and increases homothetically until it reaches the edges of the sample. Intermediate pictures of the growth are shown in Fig. 7(b) for a drilled pellet (left) and for a plain pellet (right). The square pattern is distinguishable in both cases with the seed at its centre. Note that the seeds were cut with edges parallel to the a- or b- directions, which is why they coincide with the growing domain borders.

The various square or circular-shaped Y123 were grown into a single domain including an interconnected structure. Optical macrographs of as-grown samples with holes are shown in Fig. 8. Fig. 9 (a and b) illustrate the cross sections of plain and perforated samples. The porosity is drastically reduced for the drilled sample. For the plain sample, a large porosity and crack zones are noticeable. Scanning Electron Microscopy between two holes shows (i) the compact, crack–free microstructure and (ii) a uniform distribution of fine Y211 particles into the Y123 matrix [41].

Fig. 10 presents the flux trapping obtained on plain and perforated samples (36 mm in diameter and 15 mm in height) after conventional oxygenation at 450 °C for 150 hours. The samples (Fig. 4c) were previously magnetized at 1 T, 77 K, using an Oxford Inc. superconducting coil. The 3D representation of the magnetic flux shows the single dome in the both cases corresponding to the signature of a single-domain. The network of the holes has not affected the current loops at the large scale. This result was confirmed by the neutron diffraction measurements (D1B line at ILL, France) showing [52] only one single domain bulk orientation with mean c-axes parallel to the pellet axis. The trapped field value is higher in the perforated pellet (583 mT) than in the plain one (443 mT). This represents an increase of 32% for the drilled sample compared with the plain one, in agreement with our previous report [42]. This increasing of the trapped field value is probably due to: (i) better oxygenation and/or less cracks and porosities for the drilled pellet, as illustrated by Fig. 3b, (ii) strong pinning, because the hole could be favourable to the vortices' penetration, (iii) enhancement of the cooling, because the sample with holes offers a large and favourable surface exchange into the liquid nitrogen bath.

On the other hand, pulse magnetization was used on the drilled and plain pellets. Both samples (16 mm diameter samples, 8 mm thick) were tested with a series of pulse magnetization experiments. A Helmholtz coil was used to generate a homogeneous magnetic field. The maximum amplitude of the magnetic field is 1 T and the raising time of the pulse was 1 ms while the decay time was 10 ms. After the pulse, the trapped field was mapped with a hall sensor probe at 0.5 mm above the sample. The result shows that for the application of a 1 T pulse the trapped field increases by up to 60% for a drilled pellet [44] as compared with to the plain one. This is an interesting result for such a form of new geometry, demonstrating the ability of the textured Y123 with multiple holes to trap a high field.

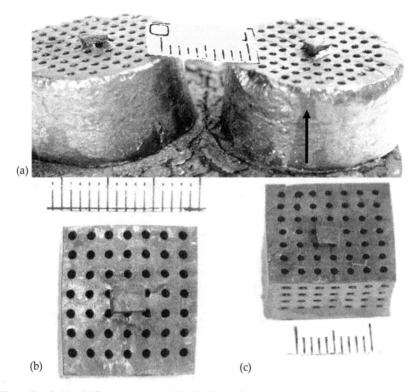

Figure 8. A batch of different as-grown drilled bulk samples (a) pellets, (b) square form and (c) interconnected samples.

Figure 9. Microstructures of the (a) thin-wall and (b) plain samples, respectively.

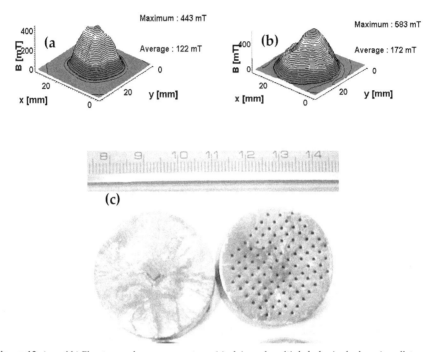

Figure 10. (a and b) Flux trapped measurements on (c) plain and multiple hole single domain pellets.

According to their thin wall geometry, the drilled bulk should be well oxygenated in comparison with the plain samples. The oxygen diffuses easily through the tube channels. The thermogravimetry technique was selected to compare the oxygenation quality of different pellets. The oxygen uptake was related to the increase of the sample weight. In this study, pellets of 16 and 24 mm diameter were used and a network of 30 holes was perforated. For each diameter, five drilled and five plain pellets were processed with the same heat treatment. All of the samples were weighted before and after the oxygenation, and the percentage of the weight gain was evaluated according to the following relation:

$$m \; (\%) = 100 \; (m_{final} - m_{initial})/m_{initial}$$

The measurements were realized twice to check reproducibility. For that, the samples after the first measurement were de-oxygenated at 900 °C, after half an hour, and followed by the quench step and then re-oxygenated. After the second measurement, the average values of the weight were estimated and plotted in Fig. 11. It was difficult to oxygenate the bulk sample with a big diameter and in this case the oxygen should diffuse into the core of the bulk. Generally, the big samples are annealed under oxygen at 400-450 °C between 150 to 600 hours [42, 46, 53, 54]. These annealing dwell times are so long in order to allow for oxygen diffusion until the core of the monolith bulk materials. The drilled samples seem to offer an advantage (a saving of time) for annealing under oxygen of the superconductor bulk. This advantage is clearly shown in Fig. 11 where 25 hours is sufficient to obtain the full oxygenated sample; in the other word, maximum weight gain is quickly achieved. In addition, thin-wall geometry was introduced to reduce the diffusion paths and to enable a progressive oxygenation strategy [54]. As a consequence, cracks are drastically reduced. In addition, the use of a high oxygen pressure (16 MPa) further speeds up the process by displacing the oxygen–temperature equilibrium towards the higher temperature of the phase diagram. The advantage of thin-wall geometry is that such an annealing can be applied directly to a much larger sample during a shorter time (72 hrs compared with 150 hrs for the plain sample). Remarkable results have been obtained by the combination of thin walls and high oxygen pressure. Fig. 13 shows the 3D distribution of the trapped flux mapping measured at 77 K on the perforated thin wall pellet. The maximum trapped field value of 0.8 T is almost twice that obtained on the plain sample (0.33 T).

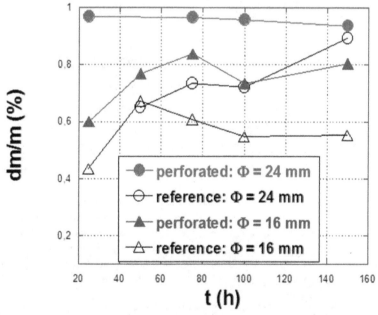

Figure 11. The influence of oxygen annealing on the oxygen uptake in the drilled and plain samples [44].

On the other hand, the effect of the number of the holes has been investigated and reported [56]. Table 1 summarizes the sample characteristics and the maximal trapped field values. We can clearly note that, for the samples having the same diameter and the same size of hole, the trapped field increases with the increase of the number of holes. An explanation could be that the better oxygenation is due to the large surface exchange with the density of the thin wall.

The Y123 domain with open holes could be reinforced, e.g. by infiltration with a low temperature melting alloy, so as to improve the mechanical properties that are useful for levitation applications or trapped field magnets. The perforated Y123 bulks with 1 or 2 mm diameter holes were dipped into the molten Sn/In alloy or an epoxy wax at 70 °C for 30 minutes in a vessel after evacuating it with a rotary pump and venting air to enable the molten alloy or liquid resin to fill up the holes. After cooling, the impregnated bulk materials were polished. Some samples were impregnated with a BiPbSnCd-alloy using the process described elsewhere [49]. Fig. 12 shows the top surface and the cross-sectional view of the impregnated Y123 bulk samples. We can see the dense and homogeneous infiltration of the wax epoxy and the Sn/In alloy. The magnetic flux mapping of the sample filled with a BiPbSnCd-alloy has been investigated. The same trapped field of 250 mT before and after impregnation has been measured. Presently, it is important to develop the specific shapes of bulk superconductors with mechanical reinforcement [52] for any practical application.

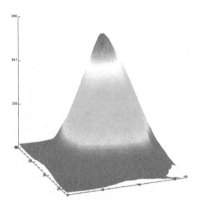

Figure 12. Flux-trapped measurements of the high pressure oxygenated thin wall sample.

sample ∅ (mm)	20.8	20.7	20.7	20.6
sample thickness (mm)	7.6	7.6	7.8	7.5
number of holes	20	37	21	85
hole ∅ (mm)	0.7	0.7	1.1	1.0
Bmax (T)	0.33	0.34	0.30	0.48

Table 1. Sample characteristics and maximal trapped field values in liquid nitrogen.

Multiple holes or porous ceramic materials, such as alumina and zirconia, are established components in a number of industrial applications such as inkjet printers, fuel injection systems, filters, structures for catalysts, elements for thermal insulation and flame barriers. The combination of a high specific surface with the ability to be reinforced in order to improve mechanical and thermal properties makes the perforated YBCO superconductors interesting candidates both for a variety of novel applications and for fundamental studies. As an example, the artificial drilled Y123 in a desired structure [43, 57] is a good candidate for resistive elements in superconducting fault current limiters (FCL) [58, 59]. In this application, the thin wall between the holes allows more efficient heat transfer between a perforated superconductor and cryogenic coolant during an over-current fault compared with conventional bulk materials. The high surface area of the perforated materials, which may be adjusted by varying the hole diameter, makes them interesting candidates for studying fundamental aspects of flux pinning, since the extent of surface pinning, and hence J_c, are expected to differ significantly from bulk YBCO grains of a similar microstructure. This new structure has great potential for many applications with improved performance in place of Y123 hole-free bulks, since it should be easier to oxygenate and to maintain at liquid nitrogen temperature during application, avoiding the occurrence of hot spot. For meandering FCL elements, cutting is a crucial step as cracks appear during this stage. This can be solved by the *in situ* zigzag shape processing of holes, as we demonstrated the feasibility of elsewhere [43].

Figure 13. Reinforcement of the drilled samples. (a) The top view of the samples filled with a BiPbSnCd-alloy, (b) with wax resin and (c) a cross-section impregnated with wax resin.

Finally, we highlight the examples among recent progress of HTS bulk applications, flywheel, power devices as motors and generators, magnetic drug delivery systems and magnetic resonance devices as well.. As shown in Fig. 14, a variety of Gd123 bulks have been tested for the employment of field pole magnets as a way of intensifying flux trapping applications. The bulk magnets are cooled down to 30 K with step-by-step pulsed-field magnetization using a homemade large dc current source. A large pulsed current is momentary applied to armature copper windings by which a pulsed magnetic field is formed and applied to the bulk field poles [60-63].

(a)   (b)

Figure 14. (a) A prototype bulk HTS motor designed for a specification of 30 kW 720 rpm and (b) a homemade pulsed-field magnetization system (TUMSAT-OLCR). This is an axial-type machine with a thermosyphon cooling system using Ne.

In summary, for the application of bulk HTS rotating machines, the enhancement of the trapped flux is a crucial task for achieving practical applications with high torque density. The increase of critical current density using artificial pinning centres marks an efficient technique for the enhancement of the properties of flux trapping. We attempted to enhance both the Jc and the trapped flux in bulk HTS with the addition of magnetic/ferromagnetic particles. An Fe-B-Si-Nb-Cr-Cu amorphous alloy was introduced into the Gd123 matrix. The melt growth of single-domain bulks with different magnetic particles was performed in air. The enhancement of the critical current density Jc at 77 K was derived in those bulks with the addition of Fe-B-Si-Nb-Cr-Cu, while the superconducting transition temperature of 93 K was not degraded significantly. The experiment of magnetic flux trapping was then conducted under static magnetic field magnetization with liquid nitrogen cooling. In the bulk with 0.4 mol% of Fe-B-Si-Nb-Cr-Cu, the integrated trapped flux exceeds over 35% compared with the one without the addition of magnetic particles. On the other hand, the addition of CoO particles resulted in a reduction of both Jc and trapped magnetic flux. The recent results indicate that the introduction of magnetic particles gives significant effect to the flux pinning's performance.

By inserting a buffer pellet with a higher T_P when compared with the matrix between the MgO seed and the bulk precursor, the lattice mismatch and low reactivity between the RE-Ba-Cu-O matrix and the MgO seed have been overcome. The undercooling and Gd211(Nd422) content for buffer pellet processing have systematically proven that the Gd-Ba-Cu-O and Nd-Ba-Cu-O bulks (16 mm in diameter) are successfully grown by this cold-seeding method. Cold-seeding melt-growth, not limited by the maximum temperature, is realized by the present new method. It was demonstrated that the texture growth can be transferred from a high-T_P pellet to a low-T_P pellet, which may be promising for extending the growth window and processing large bulk superconductors.

The single domain of Y123 bulks with multiple holes has been processed and characterized. SEM investigations have shown that the holes' presence does not hinder the domain growth. The perforated samples exhibit a single domain character evidenced by a single dome trapped-field distribution and neutron diffraction studies. This new structure has great potential for many applications, with improved performances in place of Y123 hole free bulks, since it should be easier to maintain at liquid nitrogen temperature and/or to improve thermal conductivity during application, avoiding the appearance of hot spot. It is clear that the Y123 bulks with an artificial pattern of holes are useful for evacuating porosity from the bulk and assisting the uptake the oxygen. The ability of the Y123 material with multiple holes to trap a high field has been demonstrated. Using high pressure oxygenation, the trapped field increases up to 0.8 T at 77 K for the thin wall pellet, corresponding to 50% more than the bulk material without holes. Using pulse magnetization, the trapped fields increases by up to 60% for the drilled pellet with respect to the plain one. Superconducting bulks with an artificial array of holes can be filled with metal alloys or high strength resins to improve their thermal properties without any important decrease of the hardness [50], so as to overcome the built-in stresses in levitation and quasi-permanent magnet applications. The thin wall bulks superconducting on extruded shapes for portative permanent magnets are under development for the introduction at the large scale of this innovative approach of "material by design".

Author details

Mitsuru Izumi
TUMSAT-OLCR, Tokyo, Japan

Jacques Noudem
CRISMAT/LUSAC-UNICAEN, Caen, France

Acknowledgement

The present work was supported by KAKENHI (21360425), Grant-in-Aid for Scientific Research (B) and the "Conseil Régional de Basse Normandie, France". This work was partly performed using the facilities of the Materials Design and Characterization Laboratory, Institute for Solid State Physics, University of Tokyo. The authors would like to thank Caixuan Xu, Yan Xu, Xu Kun, Keita Tsuzuki, Difan Zhou, Shogo Hara, Yufeng Zhang, Motohiro Miki, Brice Felder and Beizhan Li.

2. References

[1] Takizawa T and Murakami M Eds. (2005) Critical Current in Superconductors. Tokyo: Fuzambo International.

[2]   Matsushita T (2007) Flux Pinning in Superconductors. Berlin, Heidelberg New York: Springer.

[3]   Babu N H, Shi Y, Iida K and Cardwell D A (2005) *A practical route for the fabrication of large single-crystal (RE)–Ba–Cu–O superconductors*, Nat. Mater. 4: 476-480.

[4]   Cardwell D A, Shi Y, Babu N H, Pathak S K, Dennis A R and Iida K (2010) Top seeded melt growth of Gd–Ba–Cu–O single grain superconductors, Supercond. Sci. Technol. 23: 034008.

[5]   Cai C, Tachibana K and Fujimoto H (2000) Study on single-domain growth of Y$_{1.8}$Ba$_{2.4}$Cu$_{3.4}$O$_y$/Ag system by using Nd123/MgO thin film as seed, Supercond. Sci. Technol. 13: 698.

[6]   Oda M, Yao X, Yoshida Y and Ikuta H (2009) Melt-textured growth of (LRE)–Ba–Cu–O by a cold-seeding method using SmBa$_2$Cu$_3$O$_y$ thin film as a seed, Supercond. Sci. Technol. 22: 075012

[7]   Yao X, Nomura K, Nakamura Y, Izumi T and Shiohara Y (2002) Growth mechanism of high peritectic temperature Nd$_{1+x}$Ba$_{2-x}$Cu$_3$O$_{7-d}$ thick film on low peritectic temperature YBa$_2$Cu$_3$O$_{7-d}$ seed film by liquid phase epitax, J. Cryst. Growth 234: 611-615.

[8]   Li T, Cheng L, Yan S, Sun L, Yao X, Yoshida Y and Ikuta H (2010) Growth and superconductivity of REBCO bulk processed by a seed/buffer layer/precursor construction. Supercond, Sci. Technol. 23: 125002.

[9]   Muralidhar M, Tomita M, Suzuki K, Jirsa M, Fukumoto Y and Ishihara A (2010) A low-cost batch process for high-performance melt-textured GdBaCuO pellets, Supercond. Sci. Technol. 23: 045033.

[10] Muralidhar M, Suzuki K, Ishihara A, Jirsa M, Fujumoto Y and Tomita M (2010) Novel seeds applicable for mass processing of LRE-123 single-grain bulks, Supercond. Sci. Technol. 23: 124003.

[11] Kim C, Lee J, Park S, Jun B, Han S and Han Y (2011) Y$_2$BaCuO$_5$ buffer block as a diffusion barrier for samarium in top seeded melt growth processed YBa$_2$Cu$_3$O$_{7-y}$ superconductors using a SmBa$_2$Cu$_3$O$_{7-d}$ seed, Supercond. Sci. Technol. 24: 015008.

[12] Zhou D, Xu K, Hara S, Li B, Deng Z, Tsuzuki K and Izumi M (2012) MgO buffer-layer-induced texture growth of RE–Ba–Cu–O bulk. Supercond. Sci. Technol. 25: 025002.

[13] Xu Y, Izumi M, Tsuzuki K, Zhang Y, Xu C, Murakami M, Sakai N and Hirabayashi I (2009) Flux pinning properties in a GdBa$_2$Cu$_3$O$_{7-\delta}$ bulk superconductor with the addition of magnetic alloy particles, Supercond. Sci. Technol. 22: 095009.

[14] Xu C, Hu A, Sakai N, Izumi M and Hirabayashi I (2005) Effect of BaO$_2$ and fine Gd$_2$BaCuO$_{7-\delta}$ addition on the superconducting properties of air-processed GdBa$_2$Cu$_3$O$_{7-\delta}$, Supercond. Sci. Technol. 18: 229-233.

[15] Cardwell D A, Babu N H, Lo W and Campbell A M (2000) Processing, microstructure and irreversibility of large-grain Nd-Ba-Cu-O, Supercond. Sci. Technol. 13: 646-651.

[16] Babu, N H and Lo, W and Cardwell, D A and Shi, Y H (2000) *Fabrication and microstructure of large grain Nd-Ba-Cu-O*. Superconductor, Science and Technology, 13: 468-472.

[17] Cima M, Flemings M, Figueredo A, Nakade M, Ishii H, Brody H and Haggerty J (1992) Semisolid solidification of high temperature superconducting oxides, J. Appl. Phys. 72: 179-191.

[18] Ishii Y, Shimoyama J, Tazaki Y, Nakashima T, Horii S and Kishio K (2006) Enhanced flux pinning properties of YBa$_2$Cu$_3$O$_y$ by dilute impurity doping for CuO chain, Appl. Phys. Lett. 89: 202514-202517.

[19] Hari Babu N, Shi Y, Pathak S K, Dennis A R and Cardwell D A (2011) Developments in the processing of bulk (RE)BCO superconductors, Physica C 471: 169-178.

[20] Hari Babu N, Reddy E S, Cardwell D A, Campbell A M, Tarrant C D and Schneider K R (2003) Artificial flux pinning centers in large, single-grain (RE)-Ba-Cu-O superconductors, Appl. Phys. Lett. 83: 4806-4809.

[21] Muralidhar M, Sakai N, Jirsa M, Murakami M and Hirabayashi I (2008) Record flux pinning in melt-textured NEG-123 doped by Mo and Nb nanoparticles, Appl. Phys. Lett. 92: 162512-162515.

[22] Hu A, Xu C, Izumi M, Hirabayashi I and Ichihara M (2006) Enhanced flux pinning of air-processed $GdBa_2Cu_3O_{7-\delta}$ superconductors with addition of ZrO_2 nanoparticles, Appl. Phys. Lett. 89: 192508-192511.

[23] Xu C, Hu A, Ichihara M, Izumi M, Xu Y, Sakai N and Hirabayashi I (2009) Transition electron microscopy and atomic force microscopy observation of Air-Processed $GdBa_2Cu_3O_{7-\delta}$ superconductors doped with metal oxide nanoparticles (Metal = Zr, Zn, and Sn), Jpn. J. Appl. Phys. 48: 023002.

[24] Muralidhar M, Sakai N, Jirsa M, Koshizuka N, Murakami M, (2004) Direct observation and analysis of nanoscale precipitates in $(Sm,Eu,Gd)Ba_2Cu_3O_y$, Appl. Phys. Lett. 85: 3504-3507.

[25] Xu Y, Izumi M, Tsuzuki K, Zhang Y, Xu C, Murakami M, Sakai N, Hirabayashi I (2009) Flux pinning properties in a $GdBa_2Cu_3O_{7-\delta}$ bulk superconductor with the addition of magnetic alloy particles, Supercond. Sci. Tech. 22: 095009.

[26] Xu K, Tsuzuki K, Hara S, Zhou D, Zhang Y, Murakami M, Nishio-Hamane D, Izumi M, (2011) Microstructural and superconducting properties in single-domain Gd–Ba–Cu–O bulk superconductors with in situ formed Fe_3O_4 ferrimagnetic particles, Supercond. Sci. Tech. 24: 085001.

[27] Xu C, Hu A, Sakai N, Izumi M and Hirabayashi I, (2006) Flux pinning properties and superconductivity of Gd-123 superconductor with addition of nanosized SnO_2/ZrO_2 particles, Physica C 445: 357-360.

[28] Yamazaki Y, Akasaka T, Ogino H, Horii S, Shimoyama J and Kishio K (2009) Excellent Critical Current Properties of Dilute Sr-Doped Dy123 Melt-Solidified Bulks at Low Temperatures, IEEE Trans. Appl. Supercond. 19: 3487-3490.

[29] Xu C, Hu A, Ichihara M, Sakai N, Hirabayashi I and Izumi M, (2007) Role of ZrO_2/SnO_2 Nano-Particles on Superconducting Properties and Microstructure of Melt-Processed Gd123 Superconductors, IEEE Trans. Appl. Supercond. 17: 2980-2983.

[30] Muralidhar M, Jirsa M and Tomita M, (2010) Flux pinning and superconducting properties of melt-textured NEG-123 superconductor with TiO_2 addition, Physica C 470: 592-597.

[31] Hari Babu N, Iida K and Cardwell D A (2007) Flux pinning in melt-processed nanocomposite single-grain superconductors, Supercond. Sci. Technol. 20: S141.

[32] Nariki S, Sakai N and Murakami M (2002) Development of Gd–Ba–Cu–O bulk magnets with very high trapped magnetic field, Physica C 378-381: 631-635.

[33] Nariki S, Sakai N, Murakami M and Hirabayashi I (2004) High critical current density in RE–Ba–Cu–O bulk superconductors with very fine RE_2BaCuO_5 particles, Physica C 412-414: 557-565.

[34] Gilligns W, Sihanek A V and Moshchalkov V V (2007) Superconducting microrings as magnetic pinning centers, Appl. Phys. Lett. 91: 202510-202513.

[35] Blamire M G, Dinner R B, Wimbush S C and MacManus-Driscoll J L (2009) Critical current enhancement by Lorentz force reduction in superconductor–ferromagnet nanocomposites, Supercond. Sci. Technol. 22: 025017.

[36] Wimbush S C, Durrell J H, Bali R, Yu R, Wang H Y, Harrington S A and MacManus-Driscoll J L (2009) Practical Magnetic Pinning in YBCO, IEEE Transactions on Appl. Supercond. 19: 3148.

[37] Wimbush S C, Durrell J H, Tsai C F, Wang H, Jia Q X, Blamire M G and MacManus-Driscoll J L (2010) Enhanced critical current in YBa$_2$Cu$_3$O$_{7-\delta}$ thin films through pinning by ferromagnetic YFeO$_3$ nanoparticles, Supercond. Sci. Technol. 23: 045019.

[38] Xu K, Tsuzuki K, Hara S, Zhou D, Xu Y, Nishio-Hamane D and Izumi M, (2011) Enhanced Flux Pinning and Microstructural Study of Single-Domain Gd-Ba-Cu-O Bulk Superconductors With the Addition of Fe-Containing Alloy Particles, IEEE Trans Magnetics. 47: 4139-4142.

[39] Tsuzuki K, Hara S, Xu Y, Morita M, Teshima H, Yanagisawa O, Noudem J, Harnois C and Izumi M, (2010) Enhancement of the Critical Current Densities and Trapped Flux of Gd-Ba-Cu-O Bulk HTS Doped With Magnetic Particles, IEEE Trans. Appl. Supercond. 21: 2714-2717.

[40] Xu Y, Tsuzuki K, Hara S, Zhang Y, Kimura Y and Izumi M, (2010) Spatial variation of superconducting properties of Gd123 bulk superconductors with magnetic particles addition, Physica C 470: 1219-1223.

[41] Meslin S, Harnois C, Chateigner D, Ouladdiaf B, Chaud X, Noudem J G (2005) Perforated monodomain YBa$_2$Cu$_3$O$_{7-x}$ bulk superconductors prepared by infiltration-growth process. Journal of the European Ceramic Society 25: 2943-2946.

[42] Chaud X, Meslin S, Noudem J, Harnois C, Porcar L, Chateigner D, Tournier R (2005) Isothermal growth of large YBaCuO single domains through an artificial array of holes, J. of Crystal Growth 275: e855-e860.

[43] Noudem J, Meslin S, Harnois C, Chateigner D and Chaud X (2004) Melt textured YBa$_2$Cu$_3$O$_y$ bulks with artificially patterned holes: a new way of processing c-axis fault current limiter meanders, Supercond. Sci. Technol. 17: 931.

[44] Noudem J, Meslin S, Horvath D, Harnois C, Chateigner D, Eve S, Gomina M, Chaud X and Murakami M (2007) Fabrication of textured YBCO bulks with artificial holes, Physica C 463–465: 301–307.

[45] Chaud X, Noudem J, Prikhna T, Savchuk Y, Haanappel E, Diko P and Zhang C (2009) Flux mapping at 77 K and local measurement at lower temperature of thin-wall YBaCuO single-domain samples oxygenated under high pressure, Physica C 469: 1200–1206.

[46] Isfort D, Chaud X, Tournier R and Kapelski G (2003) Cracking and oxygenation of YBaCuO bulk superconductors: application to c-axis elements for current limitation, Physica C 390: 341.

[47] Diko P, Fuchs G and Krabbes G (2001) Influence of silver addition on cracking in melt-grown YBCO, Physica C 363: 60.

[48] Noudem J, Tarka M and Schmitz G (1999) Preparation and characterization of electrical contacts to bulk high-temperature superconductors, Inst. Phys. Conf. Ser. No 167, EUCAS, Spain, 14-17 September: 183.

[49] Tomita M and Murakami M (2003) High-temperature superconductor bulk magnets that can trap magnetic fields of over 17 tesla at 29 K, Nature 421: 517.

[50] Fuchs G, Gruss S, Verges P, Krabbes G, Muller K, Fink J and Schultz L (2002) High trapped fields in bulk YBCO encapsulated in steel tubes, Physica C 372-376: 1131.

[51] Krabbes G, Fuchs G, Schatzle P, Gruss S, Park J, Hardinghaus F, Hayn R and Drechsler S-L (2000) YBCO - monoliths with trapped fields more than 14 T and peak effect, Physica C 341-348: 2289-2292.

[52] Noudem J, Meslin S, Horvath D, Harnois C, Chateigner D, Ouladdiaf B, Eve S, Gomina M, Chaud X and Murakami M (2007) Infiltration and Top Seed Growth-Textured YBCO Bulks With Multiple Holes, J. Am. Ceram. Soc. 90: 2784-2790.

[53] Diko P, Kracunovska S, Ceniga L, Bierlich J, Zeiberger M and Gawalek W (2005) Microstructure of top seeded melt-grown YBCO bulks with holes, Supercond. Sci. Technol. 18: 1400.

[54] Nariki S, Sakai N, Murakami M (2005) Melt-processed Gd-Ba-Cu-O superconductor with trapped field of 3 T at 77 K, Supercond. Sci. Technol. 18: S126.

[55] Chaud X, Noudem J, Prikhna T, Savchuk Y, Haanappel E, Diko P and Zhang C (2009) Flux mapping at 77 K and local measurement at lower temperature of thin-wall YBaCuO single-domain samples oxygenated under high pressure, Physica C 469: 1200-1206.

[56] Haindl S, Hengstberger F, Weber H, Meslin S, Noudem J and Chaud X (2006) Hall probe mapping of melt processed superconductors with artificial holes, Supercond. Sci. Technol. 19: 108-115.

[57] Harnois C, Meslin S, Noudem J, Chateigner D, Ouladdiaf B and Chaud X (2005) Shaping of melt textured samples for fault current limiters, IEEE Trans. Appl. Supercond. 15: 3094.

[58] Tournier R, Beaugnon E, Belmont O, Chaud X, Bourgault D, Isfort D, Porcar L and Tixador P (2000) Processing of large $Y_1Ba_2Cu_3O_{7-x}$ single domains for current-limiting applications, Supercond. Sci. Technol. 13: 886.

[59] Tixador P, Obradors X, Tournier R, Puig T, Bourgault D, Granados X, Duval J, Mendoza E, Chaud X, Varesi E, Beaugnon E and Isfort D (2000) Quench in bulk HTS materials - application to the fault current limiter, Supercond. Sci. Technol. 13: 493.

[60] Tsuzuki K, Hara S, Xu Y, Morita M, Teshima H, Yanagisawa O, Noudem J, Harnois Ch. Izumi M (2011) Enhancement of the Critical Current Densities and Trapped Flux of Gd-Ba-Cu-O Bulk HTS Doped With Magnetic Particles, IEEE Trans. Appl. Supercond. 21: 2714-2717.

[61] Miki M, Felder B, Tsuzuki K, Deng Z, Shinohara N, Izumi M, Ida T and Hayakawa H (2011) Influence of AC Magnetic Field on a Rotating Machine With Gd-Bulk HTS Field-Pole Magnets, IEEE Trans. Appl, Supercond. 21: 1185-1189.

[62] Morita E, Matsuzaki H, Kimura Y, Ogata H, Izumi M, Ida T, Murakami M, Sugimoto H and Miki M (2006) Study of a new split-type magnetizing coil and pulsed field magnetization of Gd–Ba–Cu–O high-temperature superconducting bulk for rotating machinery application, Supercond. Sci. Technol. 19: 1259.

[63] Miki M, Felder B, Tsuzuki K, Xu Y, Deng Z, Izumi M, Hayakawa H, Morita M and Teshima H, (2010) Materials processing and machine applications of bulk HTS, Supercond. Sci. Technol. 23:124001.

Chapter 5

Development and Present Status of Organic Superconductors

Gunzi Saito and Yukihiro Yoshida

Additional information is available at the end of the chapter

1. Introduction

Since the first observation of superconductivity by K. Ones with a critical temperature of superconductivity (T_c) of 4.2 K on mercury (1911), many researchers have persuaded such exciting system on organic materials with vain. Even metallic behavior was hardly seen on the organic materials. Little's theoretical proposal (1964) for high T_c superconductivity ($T_c >$ 1000 K) was based on a polymer system having both a conduction path and highly polarizable pendants, which mediate the formation of Cooper pairs in the conduction path by electron-exciton coupling [1]. There are at least two inorganic polymer superconductors without doping (graphite and diamond are superconductors by doping: see Section 5), poly(sulfur nitride) $(SN)_x$ (1975, $T_c \leq 3$ K) [2] and black phosphorus (1984, $T_c \sim 6$ K at 16 GPa and 10.7 K at 29 GPa) [3], with crystalline forms. However, so far no organic polymers have been confirmed to show superconductivity which is easily destroyed by a variety of disorder. Only crystalline polymers were reported to exhibit metallic behavior: a doped polyaniline by chemical oxidation of monomers [4] and MC_{60} (Scheme 1) having linearly polymerized $C_{60}{}^{\bullet-}$ with one-dimensional (M = Rb, Cs) or three-dimensional (M = K) metallic behavior [5].

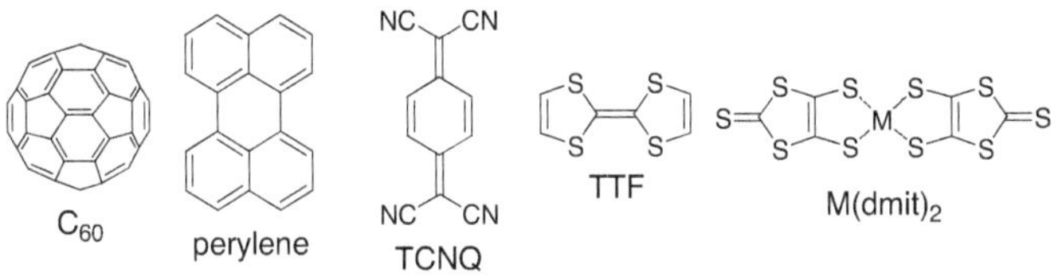

Scheme 1.

The Little's model accelerated the exploration of the conducting organic materials of low molecular weight, that had been started by the finding of highly conductive perylene•halides charge-transfer (CT) solids (10^0–10^{-3} S cm^{-1}) in early 1950s [6] and TCNQ

CT solids from the 1960s [7]. The first metallic CT solid TTF•TCNQ appeared in 1973 [8] based on the two main requirements for the conductivity, namely, (1) a uniform segregated stacking of the same kind of component molecules, and (2) the fractional CT state (uniform partial CT) of the molecules. Since TTF•TCNQ has a low-dimensional segregated stacking, it showed a metal-insulator (MI) transition (Peierls transition) below about 60 K. For TTF•TCNQ, the Peierls transition occurs by the nesting of the one-dimensional Fermi surface causing lattice distortion associated with the strong electron-phonon interaction and forms charge density wave (CDW). There are also several one-dimensional organic metals which show MI transitions by the formation of spin density wave (SDW) when the periodicity of the SDW coincides with the nesting vector of Fermi surface and no lattice distortion occurs in this case. An increase in the electronic dimensionality is inevitable to prevent the nesting of Fermi surfaces and develop superconductors. Several attempts have been made through "pressure", "heavy atom substitution", or "peripheral addition of alkylchalcogen groups" (Fig. 1). The latter two correspond to the enhancement of the self-assembling ability of the molecules.

Appropriate examples taking TTF derivatives are shown in Fig. 1. Based on TMTSF molecules several superconductors under pressure have been prepared with warped one-dimensional Fermi surface since 1980 (a in Fig. 1) [9–14]. In general, the ratio of transfer energies ($t_{//} / t_\perp$) is larger than 3 for one-dimensional Fermi surface and a closed two-dimensional Fermi surface is formed when $t_{//} \leq 3t_\perp$, where $t_{//}$ and $t_\perp$ are the transfer energies along the directions of the largest and second largest intermolecular interactions. The BO (BEDO-TTF) molecules afforded stable two-dimensional metals having two-dimensional Fermi surface (b in Fig. 1) owing to the strong self-assembling ability by intermolecular S···S and hydrogen-bonds [15], and only two superconductors are known since 1990 ($T_c \leq 1.5$ K). The substitution of an ethylenedioxy group with an ethylenedithio group (BO → ET (BEDT-TTF)) destabilized the metallic state of BO compounds and provided unstable two-dimensional conductors (c in Fig. 1). Consequently, variety of superconductors and other functional solids have been developed based on two-dimensional metals of ET since 1982 ($T_c \leq 13.4$ K) [16–20] and its analogues ($T_c \leq 10$ K) [21].

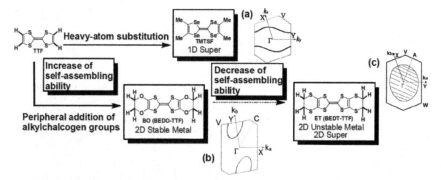

Figure 1. Strategy for chemical modification of the TTF molecule to increase (arrows) or decrease (dotted arrow) the electronic dimensionality by the aid of enhancement or suppression of the self-assembling ability of the molecules, respectively [16]. Typical Fermi surfaces of TMTSF (a: (TMTSF)$_2$NbF$_6$), BO (b: (BO)$_{2.4}$I$_3$), and ET (c: β-(ET)$_2$I$_3$) CT solids are depicted.

In this review, we mainly introduce the development and present status of organic superconductors of CT type based on electron donor molecules such as ET, electron acceptor molecules such as C_{60} (highest T_c = 38 K at 0.7 GPa), aromatic hydrocarbons (highest T_c = 33 K at ambient pressure (AP)), M(dmit)$_2$ (highest T_c = 8.4 K at 0.45 GPa $//b$) and graphite (highest T_c = 15.1 K at 7.5 GPa), carbon nanotube (T_c ~ 15 K in zeolite), and B doped diamond (T_c = 11 K at AP). Besides those, single component organic compounds show superconductivity (T_c ≤ 2.3 K at 58 GPa). The most reported T_c values of CT solids of C_{60}, aromatic hydrocarbons, those recently prepared, and those under pressure are the on-set values that are approximately 0.5–1 K higher than the mid-point T_c values. All donor based superconductors are stable in open air, however, only M(dmit)$_2$ superconductors are stable among the acceptor based superconductors.

Most of the superconducting phases of TMTSF, ET, and C_{60} materials and also of oxide superconductors reside spin-ordered phases such as SDW and antiferromagnetic (AF) phases. We briefly describe the recent development of superconductors having superconducting phase next to spin-disorder state (quantum spin liquid state).

2. Preparation of organic superconductors

CT solids are prepared mainly by the following three redox reactions: (1) electrocrystallization (galvanostatic and potentiostatic), (2) direct reaction of donors (D) and acceptors (A) in the gaseous, liquid, or solid phase, and (3) metathesis usually in solution (D•X + M•A → D•A + MX, M: cation, X: anion). In the latter two cases, single crystals are produced by the diffusion, concentration, slow cooling, or slow cosublimation methods.

Electrocrystallization (main procedures in detail and corresponding references are described in Section 11 of Ref. 17) is performed with a variety of glass cells, as shown in Fig. 2. Strictly speaking, the potentiostatic method is the proper way, in which a three-compartments cell is employed and one of the compartments contains the reference electrode, such as saturated calomel or Ag/AgCl electrode. However, this method is troublesome when a large number of crystal-growth runs are performed for a long period of time due to the following: 1) the contamination through the use of a reference electrode cell, and 2) the limited space for the experiment. The galvanostatic method is much more convenient than the potentiostatic one from these points of view. An H-cell (20 ml or 50 ml capacity) and an Erlenmeyer-type cell (100 ml) with a fine- porosity glass-frit equipped with two platinum wire electrodes (1–5 mm in diameter) have been used (Fig. 2).

There are many factors and tricks to grow single crystals of good quality. The important factors besides both the purity and the concentration of the component materials are the kinds of solvent and electrolyte, the surface of the electrode, the current (0.5–5 µA), and temperature. THF (tetrahydrofuran), CH_2Cl_2, TCE (1,1,2-trichloroethane), chlorobenzene, CH_3CN, and benzonitrile are commonly utilized solvents. The addition of 1–10 v/v% ethanol occasionally accelerates the crystal growth. As for the electrolyte, solubility in organic solvent is an important factor and usual electrolytes are tetrabutylammonium (TBA) or tetraphenylphosphonium salt of anion X. Sometimes, the electrolyte is a combination of soluble and insoluble materials. For example, single crystals of κ-(ET)$_2$Cu(NCS)$_2$ were prepared using 1) CuSCN + KSCN + 18-crown-6-ether, 2) TBA•SCN + CuSCN, or 3) Cu(NCS)$_2$•K(18-crown-6-ether). Low solubility of the components of the electrolyte in the specific solvent usually retarded single crystal growth. Ionic liquids such as 1-ethyl-3-methylimidazolium (EMI, Scheme 2) salts of X were found to afford single crystals of high quality, recently. Regarding the surfaces of electrodes each research group has special treatments such as burning (but not melting) or polishing with very fine powder. The electrode surface can be treated by applying a current to switch the polarity in a 1 M H$_2$SO$_4$ solution. When the radical species are unstable in solution, CT solids can be grown by applying a high current at very low temperatures; $e.g.$, salts of fluoranthene (–30 °C, 2 mA, Ni electrode), naphthalene, and azulene.

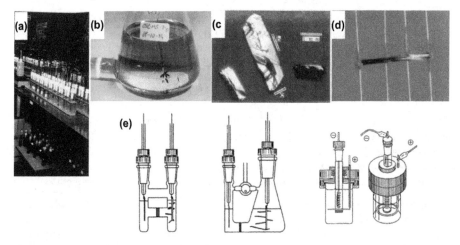

Figure 2. (a) Galvanostatic electrocrystallization using 20 ml cells on the desk. Under the desk the diffusion method is seen. (b) Single crystals of κ-(ET)$_2$Cu(NCS)$_2$ on the electrode in 100 ml cell and (c) showing two-dimensional conducting plane (bc). (d) Single crystal of (TMTSF)$_2$ClO$_4$ with four gold wires connected by gold paste. (e) Typical glass cells for electrocrystallization.

$$Me-N \overset{\textstyle\frown}{\underset{\textstyle\smile}{\oplus}} N-Et$$

EMI

Scheme 2.

Besides the electrocrystallization, superconducting single crystals of good quality were prepared by direct chemical oxidation of ET with iodine in gas or with TBA•I$_3$ or TBA•IBr$_2$ in solution. Better-quality single crystals of (TTF)[Ni(dmit)$_2$]$_2$ were obtained by the diffusion method of the metathesis reaction rather than electrocrystallization. No single crystals of the electron acceptor based superconductors were obtained except for the M(dmit)$_2$ system.

The earliest route for reductive intercalation of C_{60} solids by alkali or alkaline-earth metals is the vapor-solid reaction by vacuum annealing. Almost all superconductors based on graphite and polyaromatic hydrocarbons have been obtained in accordance with this synthetic route. Besides pristine alkali metals, sodium mercury amalgams, sodium borohydride, alkali azides, and alkali decamethylmanganocene have been utilized for source of alkali metal vapors to reduce C_{60} solids. Disproportionation reaction between C_{60} and M_6C_{60} (M: alkali metal) has been sometimes utilized to obtain superconducting M_3C_{60} or non-superconducting M_4C_{60}. For a low-temperature solution route, liquid ammonia and methylamine have been sometimes utilized for the reaction media. Especially, Cs_3C_{60} with the highest T_c among the C_{60} superconductors can be obtained only when the stoichiometric amounts of cesium metal and C_{60} were reacted in the dissolved methylamine media (see Section 3-2), while the conventional vapor-solid reaction gives energetically stable Cs_1C_{60} and Cs_4C_{60} instead of Cs_3C_{60} with nominal composition.

3. Structures and properties

3.1. Superconductors based on electron donors

3.1.1. One-dimensional superconductors (TMTSF and TMTTF families)

TMTSF [9–14] has provided eight quasi-one-dimensional superconductors; $(TMTSF)_2X$ with highest $T_c \sim 3$ K [11] (Table 1). Most of them were prepared by electrocrystallization using TBA•X as electrolyte except the NbF_6 salt which can be only prepared by using ionic liquid EMI·NbF_6 [12]. They are isostructural to each other and the crystal structure of $(TMTSF)_2NbF_6$ is depicted in Fig. 3, where TMTSF molecules form a zigzag dimer that forms a segregated column along the face-to-face direction (a-axis) with no short Se···Se atomic contacts (Fig. 3a,3b). Along the side-by-side direction (b-axis), rather short Se···Se atomic contacts were seen (Fig. 3c), however, those less than the sum of the van der Waals radii (3.80 Å) are present only for X = ClO_4 and FSO_3. For the PF_6 salt, t_a and t_b were estimated to be 0.25–0.30 eV and 0.031 eV, respectively. Consequently, the Fermi surface of $(TMTSF)_2X$ is not closed, but open with fair warping due to the lack of adequate side-by-side transfer interactions (Fig. 1a).

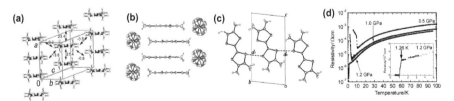

Figure 3. Crystal structure of $(TMTSF)_2NbF_6$ [12]. (a) Segregated column of TMTSF molecules. The numbers indicate the overlap integrals in 10^{-3} units. (b) Zigzag stacking of the TMTSF column. (c) Se···Se atomic contacts (d_7, d_9) along the side-by-side direction. (d) Temperature dependence of resistivity under pressure.

Salts with octahedral anions exhibited MI transition at 11–17 K at AP due to SDW, and a superconductivity appeared with on-set T_c of ca. 1 K at 0.6–1.2 GPa (Fig. 3d). Salts with (pseudo)tetrahedral anions, on the other hand, exhibited order-disorder (OD) transition of anion molecules when the superlattice created was coincide with the nesting vector of the warped Fermi surface ($2a \times 2b$, Fig. 1a).

The isomorphous (TMTTF)$_2$X (TMTTF: Scheme 3) salts displayed superconductivity under high pressure of 2.6–9 GPa with T_c less than 3 K for X = Br, BF$_4$, PF$_6$, and SbF$_6$ [22–25]. Table 1 summarizes superconductors of (TMTSF)$_2$X and (TMTTF)$_2$X showing σ_{RT}, temperature at which conductivity shows maximum (T_{max}) due to an MI transition, critical pressure to induce superconductivity (P_c), T_c, phenomena which cause the MI transition (transition temperature), and superlattice after the ordering of the anion molecules.

(TMTSF)$_2$ClO$_4$ is the only AP superconductor among them and shows no Hebel-Slichter coherence peak [26], which should be observed just below T_c for a normal BCS-type superconductor having an isotropic gap [27], in the early measurements of relaxation rate of ^{1}H NMR absorption. Later, the thermal conductivity suggested a fully-gapped order parameter [28]. The superconducting coherent lengths are 710 ($//a$), 340 ($//b$), and 20 Å ($//c$), indicating a quasi-one-dimensional character. The application of magnetic field breaks the superconducting state and induces a sequence of SDW (field-induced SDW: FISDW) states above 3 T. Upper critical magnetic field H_{c2} of the PF$_6$ salt (H_{c2} = 6 ($//b'$), 4 ($//a$) T at 0.1 K) is far beyond the Pauli limit (H_{Pauli}) for the BCS-type superconductor with weak coupling [10]. A generalized phase diagram including (TMTSF)$_2$X and (TMTTF)$_2$X indicates that the superconducting phase neighbors the magnetic SDW phase (Fig. 4) [29].

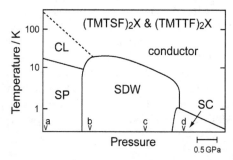

Figure 4. Generalized phase diagram for the (TMTSF)$_2$X and (TMTTF)$_2$X [29]. CL, SP, and SC refer to charge-localized (which corresponds to charge-ordered state), spin-Peierls, and superconducting states, respectively. The salts **a–d** at AP locates in the generalized diagram. **a**: (TMTTF)$_2$PF$_6$, **b**: (TMTTF)$_2$Br, **c**: (TMTSF)$_2$PF$_6$, **d**: (TMTSF)$_2$ClO$_4$.

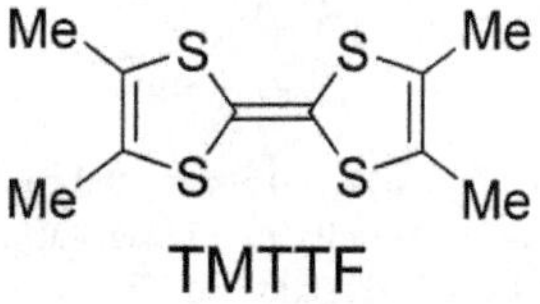

TMTTF

Scheme 3.

X	Symmetry	σ_{RT} / S cm^{-1}	T_{max}[a) / K	P_c / GPa	T_c[b) / K	Characteristics[c)
TMTSF system						
PF$_6$	octahedral	540	12-15	0.65	1.1	SDW (12 K), FISDW
AsF$_6$	octahedral	430	12-15	0.95	1.1	SDW (12 K, J = 604 K)
SbF$_6$	octahedral	500	12-17	1.05	0.38	SDW (17 K)
NbF$_6$	octahedral	120	12	1.2	1.26	SDW (12 K)
TaF$_6$	octahedral	300	15	1.1	1.35	SDW (11 K)
ClO$_4$	tetrahedral	700	–	0	1.4	OD (24 K, $a \times 2b \times 2c$), FISDW, (γ= 10.5, β = 11.4, Θ = 213)
		–	5	–	–	SDW (5 K) by rapid cool
ReO$_4$	tetrahedral	300	~182	0.95	1.2	OD (177 K, $a \times 2b \times 2c$)
FSO$_3$	pseudo-tetrahedral	1000	~88	0.5	3	OD (88 K, $a \times 2b \times 2c$)
TMTTF system						
PF$_6$	octahedral	20	245	5.2-5.4	1.4-1.8	spin-Peierls (15 K)
SbF$_6$	octahedral	8	150	5.4-9	2.8	charge-order, AF (8 K)
BF$_4$	tetrahedral	50	190	3.35-3.75	1.38	OD (40 K), SDW and SC (coexist)
Br	spherical	260	100	2.6	1.0	AF (15 K)

a) T_{max}: temperature at maximum conductivity. b) T_c: on-set. c) SDW: spin density wave, OD: order-disorder transition of anion and newly formed superlattice. FISDW: field-induced SDW. γ and β are important quantities experimentally determined to obtain $D(\varepsilon)$ (eq. 1) and Θ (eq. 2) which are related with T_c by eq. 3 for the BCS-type superconductors.

Table 1. Organic superconductors of (TMTSF)$_2$X and (TMTTF)$_2$X

γ: Sommerfeld coefficient, mJ mol^{-1} K^{-2}

$$\gamma = \pi^2 k_B^2 D(\varepsilon_F)/3 \tag{1}$$

β: mJ mol^{-1} K^{-4}

$$\beta = 48\pi N k_B/5\Theta^3 \tag{2}$$

Θ: Debye temperature, K

$$T_c \propto \Theta \exp(-1/V_{el-ph}D(\varepsilon_F)) \tag{3}$$

k_B: Boltzmann constant, g: coupling constant, V_{el-ph}: electron-phonon coupling potential, $D(\varepsilon_F)$: density of states at Fermi level per one spin.

3.1.2. Two-dimensional superconductors (BO, ET, and BETS families)

3.1.2.1. BO superconductors

TTF derivatives with "peripheral addition of alkylchalcogeno groups" were found to be effective to increase dimensionality of CT solids and suppress the Peierls-type MI transition for many BO [15] and ET [16–20] conductors. The robust intermolecular interactions in the BO complexes have provided a metallic state even in the strongly disordered systems. The strong two-dimensionality in the BO complexes hardly exhibited any phase transition including the superconductivity (only two superconductors with $T_c \leq 1.5$ K were found) [30,31].

3.1.2.2. ET superconductors

The first ET two-dimensional organic metal down to low temperatures is $(ET)_2(ClO_4)(TCE)$ [32]. Since then, hundreds of ET solids have been prepared. ET molecules tend to pile up one after the other with sliding to each other so as to minimize the steric hindrance caused by the terminal ethylene groups. A neutral ET molecule is non-planar and becomes almost flat on formation of the partial CT complex except the terminal ethylene groups which are thermally disordered at high temperatures. Segregated packing of such molecules leaves cavities along the molecular long axis, where counter anions and sometimes solvent molecules occupy. It was pointed out that the ethylene conformation is one of the key parameters determining the physical and structural properties including the superconductivity [33]. ET molecule also has a strong tendency to form proximate intermolecular S···S contacts along the side-by-side direction leading to an increment of the side-by-side transfer integrals $t_\perp$ (Fig. 5). The ET conductors are composed of alternating structures of two-dimensional conducting layer and insulating anion layer. Significant donor···anion interactions arise from the short atomic contacts between the ethylene hydrogen atoms of ET and anion atoms around the anion openings in the anion layer as schematically shown in Fig. 5b [34].

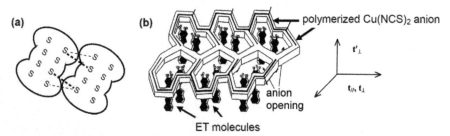

Figure 5. (a) Schematic figure of an example of S···S atomic contacts observed in ET salts. Thick dotted lines: S_{in}···S_{out}. Thin dotted lines: S_{out}···S_{out}. (b). Schematic view of κ-$(ET)_2Cu(NCS)_2$ indicating anion openings and transfer interactions ($t_{//}$, $t_\perp$ and $t'_\perp$) [34]. ET dimers are nearly orthogonally aligned (κ-type packing). For κ-type salt, $t_{//} \sim t_\perp \gg t'_\perp$.

The steric hindrance exerted by bulky six-membered rings of ET molecules prevents the formation of intermolecular S_{in}···S_{in} contacts (S_{in}: sulfur atom in the TTF skeleton, Fig. 5a). No particular patterns of intermolecular S_{in}···S_{out} (S_{out}: sulfur atom in the six-membered ring) are favorable as well. As a consequence various kinds of S···S contacts are produced depending on the donor packing patterns (α-, β-, θ-, κ-phases, and so forth, see Section 3-1-2-5), and are comparable to the other intermolecular interactions; i.e., face-to-face (π-π) and donor···anion interactions. Any interactions could not solely determine the donor packing picture. It is thus much more difficult to predict the donor packing pattern for the ET system compared to those for TMTSF, especially for salts with small and discrete anions such as I_3, ClO_4, PF_6, etc., where polymorphic isomers are frequently afforded. In the salts with discrete linear anions such as I_3 and I_2Br, the component molecules have great freedom of motion and the donor packing pattern can be changed by thermal or pressure treatment [35,36]. Scanning tunneling microscope (STM) measurements [37–39] revealed that the surface structure of β-$(ET)_2I_3$ crystals contains many defects, voids, and reconstruction of donor packing attributed to the unstable structure of the anion layers, while the surface structures of salts with polymerized anions such as κ-$(ET)_2Cu(NCS)_2$ (Fig. 7b in Section 3-1-2-3) and α-$(ET)_2MHg(SCN)_4$ (M = NH_4 and K) are stable with no defects.

The polymerized anions in κ-$(ET)_2Cu(NCS)_2$ form the insulating layer having openings as seen in Fig. 5b. Two ET molecules form a dimer unit which fits into each opening. In more accurate description, the hydrogen atom of one ethylene group of ET molecule fits into the core created by anion molecules, like a key-keyhole relation. The position of such an ethylene hydrogen atom projected onto the anion cores produces unique patterns; called α-type (5 superconductors), β-type (6 superconductors), θ-type (1 superconductor), and κ-type (about 31 superconductors) [40]. It means that the ET molecules arrange according to the anion core or opening pattern created by polymerized anions.

Different kinds of ET$\cdots$ET (π-π, S$\cdots$S) and ET$\cdots$anion (hydrogen bonds) intermolecular interactions, large conformational freedom of ethylene groups, flexible molecular framework, fairly narrow bandwidth (W), and strong electron correlations, which are represented by on-site Coulomb repulsion energy U, gave a rich variety of complexes with different crystal and electronic structures ranging from insulators to superconductors (corresponding references are cited in Ref. 16): Mott insulators (including spin-Peierls systems, antiferromagnets, spin-ladder systems, and quantum spin liquid), one-dimensional metals with CDW transition, two-dimensional metals with CDW transition, two-dimensional metals with FISDW transition, charge-ordered insulators, monotropic complex isomers, and two-dimensional metals down to low temperatures.

About 60 ET superconductors have so far been known. Table 2 summarizes selected ET superconductors and related salts. They are classified into three classes based on the transport behavior at AP: 1) Salts in Class I are metallic down to rather low T_c. 2) Salts in Class II are close to a Mott insulator and a poor metal showing $T_c > 10$ K. 3) Salts in Class III are insulators (Mott, CDW, or charge order). Figure 6 compares the temperature dependence of resistivity for several κ-$(ET)_2X$ with that of β-$(ET)_2AuI_2$ (6, Class I) which exhibited metallic behavior down to T_c at 4.9 K [48]. κ-$(ET)_2Cu(CN)[N(CN)_2]$ [49] (1, Class II) showed a monotonous decrease of resistivity with upper curvature down to T_c. κ-$(ET)_2Cu[N(CN)_2]Br$ [55] (3, Class II) exhibited similar behavior to that of κ-$(ET)_2Cu(NCS)_2$ [51] (2, Class II) except a metallic regime near RT in 2. They have a semiconductive region down to 70–80 K followed by a metallic behavior down to T_c. κ-$(ET)_2Cu[N(CN)_2]Cl$ [57–60] (4, Class III) ia a Mott insulator and showed a semiconductor (ε_g = 24 meV)-semiconductor (ε_g = 104 meV) transition at ca. 42 K due to an AF fluctuation resulting in a weak ferromagnet below 27 K (Néel temperature T_N = 27 K). Under a low pressure, it showed a similar temperature dependence to that of κ-$(ET)_2Cu[N(CN)_2]Br$. κ-$(ET)_2Cu_2(CN)_3$ [61–65] (5,

Class III) is semiconductive (a Mott insulator) and under pressure it also behaves similarly to **4** (semiconductor-metal-superconductor).

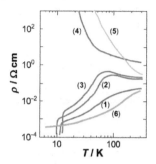

Figure 6. Temperature dependences of resistivity of 10 K class superconductors κ-(ET)$_2$Cu(CN)[N(CN)$_2$] (**1**), κ-(ET)$_2$Cu(NCS)$_2$ (**2**), κ-(ET)$_2$Cu[N(CN)$_2$]Br (**3**), and κ-(ET)$_2$Cu[N(CN)$_2$]Cl (**4**), which are compared with that of a good metal with low T_c β-(ET)$_2$AuI$_2$ (**6**) and a Mott insulator κ-(ET)$_2$Cu$_2$(CN)$_3$ (**5**) at AP.

Class, Salt	CuL$_1$L$_2$ L$_1$, L$_2$	σ_{RT}/ S cm^{-1}	$T_c^{a)}$ / K H-salt	D-salt	t'/t	U/W	Ground state at AP[b)]	Characteristics, Quantum oscillations	Ref
I κ-(ET)$_2$I$_3$		30	3.6*	–	0.58		SC	$\gamma = 18.9$, $\beta = 10.3$, $\Theta = 218$, SdH, dHvA	[41]
β-(ET)$_2$I$_3$		60	1.5, 2.0, 8.1	–	0.55		SC	SdH, dHvA, AMRO	[42–47]
6 β-(ET)$_2$AuI$_2$		20–60	4.9	–	–		SC	SdH, dHvA	[48]
II 1 κ-(ET)$_2$Cu(CN)[N(CN)$_2$]	CN, N(CN)$_2$	5–50	11.2*	12.3*	0.66–0.71	0.87	SC	AMRO	[49,50]
2 κ-(ET)$_2$Cu(NCS)$_2$	SCN, NCS	5–40	10.4*	11.2*	0.82–0.86	0.94	SC	$\gamma = 25$, $\beta = 11.2$, $\Theta = 215$, SdH, dHvA, AMRO	[51–54]
3 κ-(ET)$_2$Cu[N(CN)$_2$]Br	N(CN)$_2$, Br	5–50	11.8*	11.2*	0.68	0.92	SC	$\gamma = 22$, $\beta = 12.8$, $\Theta = 210$, SdH, AMRO	[55,56]
III 4 κ-(ET)$_2$Cu[N(CN)$_2$]Cl	N(CN)$_2$, Cl	2	12.8/0.0 3 GPa	13.1	0.75	0.90	AF	SdH, AMRO, $T_N = 27$ K	[57–60]
5 κ-(ET)$_2$Cu$_2$(CN)$_3$	CN, CN/NC	2–7	6.8– 7.3**		1.06	0.9	SL	SdH, AMRO	[49,61–65]
ET•TCNQ		10			–		AF	$T_N = 3$ K	[66]
β'-(ET)$_2$ICl$_2$		3×10^{-2}	14.2/8.2 GPa		–		AF	$T_N = 22$ K	[67,68]
β'-(ET)$_2$AuCl$_2$		~10^{-1}			–		AF	$T_N = 28$ K	[68]
β'-(ET)$_2$BrICl		~10^{-2}	7.2/8.0 GPa				AF	$T_N = 19.5$ K	[69]

a) *: mid-point. **on-set under uniaxial strain (see Fig. 9a). Others are the on-set values under hydrostatic pressure. b) SC: superconductor, SL: spin liquid.

Table 2. Selected ET conductors and superconductors. Except ET•TCNQ, the compound is represented by *Greek alphabet*-(ET)$_2$X (Greek alphabet: type of donor stacking, L$_1$, L$_2$: ligand). a) Class **I** : good metal with low T_c, **II**: 10 K class AP superconductor, **III**: Mott insulator. **1–6** are the numbers in Fig. 6.

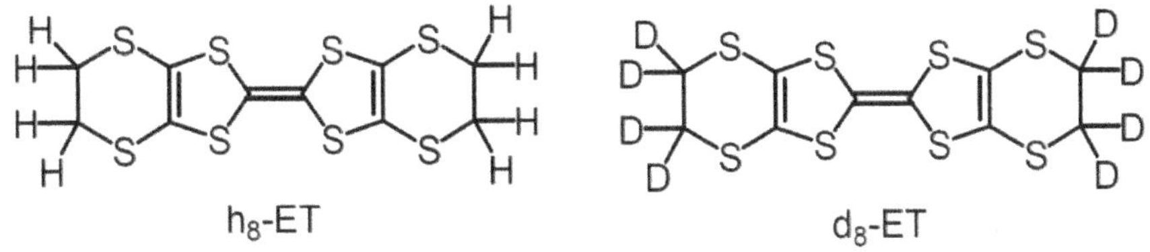

h₈-ET   d₈-ET

Scheme 4.

Table 2 summarizes T_c of H- and D-salts (salt using h₈-ET and d₈-ET, respectively; Scheme 4). The calculated U/W values close to unity suggest that those salts have strong electron correlation. Currently β'-(h₈-ET)₂ICl₂ (on-set T_c = 14.2 K at 8.2 GPa [68], mid-point T_c of 13.4 K is estimated) and D-salt of **4** (T_c = 13.1 K at 0.03 GPa) [58] show the highest T_c under pressure, while both are Mott insulators at AP. At AP, D-salt of **1** shows the highest T_c of 12.3 K [50] followed by H-salt of **3** (T_c = 11.8 K) [55,56]. These salts are electronically clean metals as evidenced by the observation of quantum oscillations (Shubnikov-de Haas (SdH), de Haas-van Alphen (dHvA)), and geometrical oscillations: angular dependent magnetoresistance oscillation (AMRO), which afford topological information for the Fermi surface [19,20,70].

Next, we will focus mainly on the κ-type superconductors with polymerized anions.

3.1.2.3. κ-type ET conductors

The κ-type superconductors κ-(ET)₂CuL₁L₂ (**1–5** in Table 2) share some common structural and physical properties [49–65]. Figure 7 shows the crystal structure of the prototype H-salt of **2**, anion structures, donor packing pattern, and calculated Fermi surface [51–54]. Table 2 summarizes the two kinds of ligand (L₁, L₂) in a salt and ratio t'/t for triangle geometry of ET dimers. These salts have polymerized anions in which ligand L₁ forms infinite chain by coordinating to Cu^{1+} and ligand L₂ coordinates to Cu^{1+} as pendant. ET molecules form a dimer and the ET dimers are arranged nearly orthogonally to each other forming two-dimensional conducting ET layer in the bc-plane which is sandwiched by the insulating anion layers along the a-axis (Fig. 7a). Cu^{1+} and SCN form Cu···SCN···Cu···SCN··· zigzag infinite chain along the b-axis and other ligand SCN coordinates to Cu^{1+} by N atom to make an open space (indicated by ellipsoid in Fig. 7b, also see Fig. 5) to which an ET dimer fits. An ET dimer has one spin, and the dimers form triangle (Fig. 7c) whose shape is represented by the ratio t'/t (Fig. 7d).

H-salt of **4** showed a complicated T-P phase diagram (Fig. 8a) [72–76]. Thoroughgoing studies under pressure showed a firm evidence of the coexistence of superconducting (**I-SC-2** phase: **I-SC** = incomplete superconducting) and AF phases [72–78], where the radical electrons of ET molecules played both roles of localized and itinerant ones. Under a pressure of ca. 20–30 MPa another incomplete superconducting phase (**I-SC-1**) appeared and the complete superconducting (**C-SC**) phase neighbored to this phase at higher pressures. Below these superconducting phases, reentrant nonmetallic (**RN**) phase was observed. Similar T-P phase diagrams were obtained for κ-(d₈-ET)₂X (X = Cu[N(CN)₂]Cl and

Cu[N(CN)$_2$]Br) with a parallel shift of pressure. They occur at the higher and lower pressure sides of the κ-(h$_8$-ET)$_2$Cu[N(CN)$_2$]Cl for the Br and Cl salts, respectively. Contrary to the H-salt, κ-(d$_8$-ET)$_2$Cu[N(CN)$_2$]Cl exhibited no coexistence of the superconducting and AF phases. At AP, κ-(ET)$_2$Cu[N(CN)$_2$]I is a semiconductive and becomes superconducting under hydrostatic pressure above 0.12 GPa with T_c of 7.7 K (on-set T_c = 8.2 K), though the magnetic ordering was not clarified at AP [79].

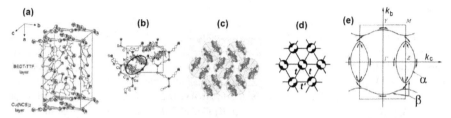

Figure 7. κ-(ET)$_2$Cu(NCS)$_2$: (a) Crystal structure. (b) Anion layer viewed along the *a*-axis has anion openings (indicated by ellipsoid) to which an ET dimer fits. Picture is the dextrorotatory form. (c) Packing pattern (κ-type) of ET dimers along the *a*-axis. (d) Schematic view of triangular lattice of ET dimers which has one spin. White and black circles represent ET molecule and ET dimer, respectively. The t'/t represents the shape of the triangle. (e) Calculated Fermi surfaces of the $P2_1$ salts (κ-(ET)$_2$Cu(NCS)$_2$, κ-(ET)$_2$Cu(CN)[N(CN)$_2$]) showed the certain energy gap between a one-dimensional electron like Fermi surface (//k_c) and a two-dimensional cylindrical hole-like one (α-orbit), while such a gap is absent in the *Pnma* salts (κ-(ET)$_2$Cu[N(CN)$_2$]Br, κ-(ET)$_2$Cu[N(CN)$_2$]Cl). For κ-(ET)$_2$Cu(NCS)$_2$, electrons move along the closed ellipsoid (α-orbit) to exhibit SdH oscillations [53], and at higher magnetic field (> 20 T) electrons hop from the ellipsoid to open Fermi surface to show circular trajectory (β-orbit, magnetic breakdown oscillations) [71].

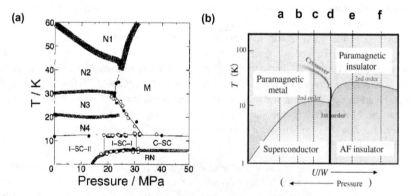

Figure 8. (a) Phase diagram of κ-(h$_8$-ET)$_2$Cu[N(CN)$_2$]Cl determined from electrical conductivity and magnetic measurements [72–76]. N1–N4: non-metallic phases, M: metallic phase, RN: reentrant non-metallic phase, I-SC-I and I-SC-II: incomplete superconducting phases. N3 shows growth of three-dimensional AF ordered phase. N4 is a weak ferromagnetic phase. (b) Proposed simplified phase diagram [80]. **a**: β-(ET)$_2$I$_3$, **b**: κ-(ET)$_2$Cu(NCS)$_2$, **c**: H-salt of κ-(ET)$_2$Cu[N(CN)$_2$]Br, **d**: D-salt of κ-(ET)$_2$[N(CN)$_2$]Br, **e**: κ-(ET)$_2$[N(CN)$_2$]Cl, **f**: β-(ET)$_2$ICl$_2$.

Alternating mixed donor packing motifs were observed in two phases of α'-κ-(ET)$_2$Ag(CF$_3$)$_4$(TCE), T_c (on-set) of which are 9.5 K and 11.0 K for the phase having two-layered (α' + κ) and four-layered (α' + κ_1 + α' + κ_2) phase, respectively [81,82]. Since α'-packing generally imparts semiconducting state, both systems have nano-scale hetero junction of semiconductive/superconductive interface, which is thought to give higher T_c in these systems in comparison with κ_1-(ET)$_2$Cu(CF$_3$)$_4$(TCE) (T_c = 4.0 K). If this explanation is correct, this is an example of interface superconductivity [83,84].

With increasing the distance between the ET dimers in Fig. 7a–d, the transfer interactions between ET dimers decrease; this may correspond to the decrease of W and to increase of $D(\varepsilon_F)$, and consequently T_c is expected to increase. According to this line of thought, higher T_c is expected for the salt having a larger anion spacing. Such a κ-type salt may be found near the border between poor metals and Mott insulators. It is true not only for κ-type but also for other types, e.g., β'-(ET)$_2$X (X = ICl$_2$ [67,68] and BrICl [69]) having high T_c in the ET family are Mott insulators at AP.

Topological structures of their Fermi surfaces studied by SdH, dHvA, and AMRO (Table 2) [19,20,53,70,71,85–87] show that the area of the closed Fermi surface relative to the first Brillouin zone and cyclotron mass calculated from SdH oscillations are 15.7% (α-orbit in Fig. 7e, 3.5m_e) and 105% (β-orbit in Fig. 7e, 6.5m_e) for **2** at AP, 4.4% (0.95m_e) at 0.9 GPa and *ca.* 100% (6.7m_e) at AP for **3**, and 15.5% (1.7m_e) and 102% (3.5m_e) at 0.6 GPa for **4**. Fermi surface of **2** (Fig. 7e) calculated based on the crystal structure is in good agreement with these observed data.

The followings are the superconducting characteristics of κ-type ET superconductors, some of which differ from those of the conventional BCS superconductors.

1. Upper critical magnetic field H_{c2}: **2** gave higher H_{c2} values for the magnetic field parallel to the two-dimensional plane than H_{Pauli} based on a simple BCS model [88,89].
2. Coherent length ξ. The superconducting coherent lengths are 29 and 3.1 Å for **2** at 0.5 K along the two-dimensional plane ($\xi_{//}$) and perpendicular to the plane ($\xi_\perp$). The $\xi_{//}$ is larger than the lattice constants, however, the $\xi_\perp$ is much smaller than the lattice constant indicating the conducting layers along this direction is Josephson coupled.
3. Symmetry of superconducting state: No Hebel-Slichter coherence peak was observed in both **2** and **3** in ^{1}H NMR measurements, ruling out the BCS s-wave state. The symmetry of the superconducting state of **2** had been controversially described as normal BCS-type or non-BCS type, however, STM spectroscopy showed the d-wave symmetry with line nodes along the direction near $\pi/4$ from the k_a- and k_c-axes ($d_{x^2-y^2}$) [90], and thermal conductivity measurements were consistent with that [91]. STM on **3** also showed the same symmetry [92]. A recent specific heat measuremen on **2** and **3** was consistent with these results [93].
4. Inverse isotope effect: Inverse isotope effect has so far been observed for **1** [50], **2** [54], and **4** [58], while normal isotope effect for **3** [56]. The reason of the observed isotope effects is not fully understood yet consistently.

5. A very simplified T-P phase diagram for κ-(ET)$_2$X was proposed (Fig. 8b), where only the parameter U/W is taken into account. Fig. 8b includes the salts **2, 3, 4**, β-(ET)$_2$I$_3$, and β'-(ET)$_2$ICl$_2$ [80], however, the metallic behavior of **2** above 270 K and that of **1**, whole behavior of **5**, and low-temperature reentrant behavior of **3** and **4** (Fig. 8a) cannot be allocated in this diagram. The β-(ET)$_2$I$_3$ in Fig. 8b should be β_H-phase (T_c ~ 8 K, see Section 3-1-2-5) and other two β-(ET)$_2$I$_3$ salts of T_c ~ 1.5 K and ~2 K phases cannot be allocated though they should have the same U/W values. T_c of **4** is higher than that of β-(ET)$_2$ICl$_2$ in Fig. 8b contrary to the experimentally observed T_c results. This phase diagram and "geometrical isotope effect" [94] point out that T_c's of β-(ET)$_2$I$_3$, **2**, and **3** decrease with increasing pressure if only the parameter U/W or $D(\varepsilon_F)$ is taken into account. This tendency has been observed under hydrostatic pressure but not under uniaxial pressure (see Sections 3-1-2-4, 3-1-2-5, 3-1-3). Thus the phase diagram in Fig. 8b remains incomplete, despite it is frequently used to explain the general trends for these salts.

3.1.2.4. Quantum spin liquid state in κ-(ET)$_2$Cu$_2$(CN)$_3$ and neighboring superconductivity

As mentioned, κ-type packing is characterized by the triangular spin-lattice (Fig. 7c,7d) where an ET dimer is a unit with $S = 1/2$ spin [95,96]. The line shape of ^{1}H NMR absorption of κ-(ET)$_2$Cu[N(CN)$_2$]Cl [60] exhibited a drastic change below 27 K owing to the formation of three-dimensional AF ordering. On the other hand, the absorption band of κ-(ET)$_2$Cu$_2$(CN)$_3$ remains almost invariant down to 32 mK indicating non-spin-ordered state: quantum spin liquid state [63]. The appearance of spin liquid state in κ-(ET)$_2$Cu$_2$(CN)$_3$ is the consequence of significant spin frustration in this salt ($t'/t = 1.06$) in comparison with the less frustrated AF state in κ-(ET)$_2$Cu[N(CN)$_2$]Cl ($t'/t = 0.75$).

It has long been predicted that the geometrical spin frustration of antiferromagnets caused by the spin correlation in particular spin geometry (triangle, tetrahedral, Kagome (Scheme 5), etc.) prevents the permanent ordering of spins. So the spins of Ising system with AF interaction in the equilateral triangle spin lattice will not show any long-range order even at 0 K, and hence the phase, namely quantum spin liquid phase, has high degeneracy [97]. Such spin liquid state has only been predicted theoretically [98], and a variety of ideal materials have been designed and examined for long [99–102]. Since the discovery of the spin liquid state in κ-(ET)$_2$Cu$_2$(CN)$_3$, several materials have been reported to have such exotic spin state: EtMe$_3$Sb[Pd(dmit)$_2$]$_2$ [103], ZnCu$_3$(OH)$_6$Cl$_2$ [104,105], Na$_4$Ir$_3$O$_8$ [106], and BaCu$_3$V$_2$O$_8$(OH)$_2$ [107,108]. Some inorganic materials reported as spin liquid candidates were eliminated owing to the spin ordering at extremely low temperatures, etc [109–113]. Na$_4$Ir$_3$O$_8$ and two organic compounds (κ-(ET)$_2$Cu$_2$(CN)$_3$, EtMe$_3$Sb[Pd(dmit)$_2$]$_2$) may be recognized as "soft" Mott insulators and have metallic state under pressure. Only κ-(ET)$_2$Cu$_2$(CN)$_3$ has the superconducting phase next to spin-liquid state so far as described below.

The phase diagrams of TMTSF (Fig. 4), ET (Fig. 8), C$_{60}$ [114] families and also electron-correlated cuprate and iron pnictide high T_c systems [115] indicate that a magnetic ordered state (SDW, AF) is allocated next to the superconducting state. Figure 9a shows the T-P phase diagram of κ-(ET)$_2$Cu$_2$(CN)$_3$ at low temperature region by applying uniaxial strain along c- (t'/t decreases in this direction) and b- (t'/t increases in this direction) axes with epoxy-method [62,65]. In both cases, a superconducting state readily appeared nearly above 0.1 GPa since the t'/t deviates from unity; *i.e.*, strong spin frustration was released. It is very noteworthy that the spin liquid phase is neighboring to the superconducting state and its T_c is fairly anisotropic as shown in Fig. 9b. A plot of T_c vs. T_{IM}, which is a Mott insulator-metal transition temperature, indicates that in comparison with the hydrostatic pressure results, the uniaxial method afforded: 1) a much higher T_c value, 2) an increase of T_c at the initial pressure region, 3) an anisotropic pressure dependence, and 4) superconducting phase remains at higher pressure. The uniaxial strain experiments including other κ-type superconductors clearly revealed that the T_c increased as the U/W approaches unity and as the t'/t departs from unity [116].

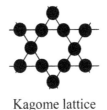

Kagome lattice

Scheme 5.

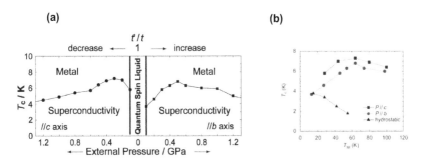

Figure 9. a) Temperature-uniaxial pressure phase diagram in the low temperature region of κ-(ET)$_2$Cu$_2$(CN)$_3$ [62,65]. The strain along the c-axis corresponds to a decrease of t'/t (left side), while the strain along the b-axis increases t'/t (right side). b) Pressure dependence of on-set T_c by the uniaxial strain and hydrostatic pressure methods. T_{IM}: a Mott insulator-metal transition temperature [65].

3.1.2.5. Other ET superconductors

One of the most intriguing ET superconductors is the salt with I$_3$ anion, which afforded α-, α_t-, β_L-, β_H-, δ-, ε-, γ-, θ-, and κ-type salts with different crystal and electronic structures. Among them, α-, α_t-, β_L-, β_H-, γ-, θ-, and κ-type salts are superconductors with T_c = 7.2, ~8, 1.5, 8.1, 2.5, 3.6, and 3.6 K, respectively [36,41,42,44,45,117–120]. The β_L-salt was converted to the β_H-salt by pressurizing (hydrostatic pressure) above 0.04 GPa and then by depressurizing while keeping the sample below 125 K [45,46]. The β_H-salt returned to β_L-salt when the salt was kept above 125 K at AP. The β_L-salt is characterized by having a superlattice appearing at 175 K with incommensurate modulations of ET and I_3 to each other [121]. The formation of the superlattice was suppressed by the pressure above 0.04 GPa. Then the two ethylene groups in an ET molecule were fixed in the eclipsed conformation to give rise to more than 5 times higher T_c in β_H-salt. The T_c of β_H-salt decreased with increasing hydrostatic pressure monotonously, however, under the uniaxial stress the further T_c increase taking a maximum at a piston pressure of 0.3–0.4 GPa is observed for both directions parallel and perpendicular to the donor stack [120]. The superconducting coherent lengths are $\xi_{//} = 630$ (//a) – 610 Å (//b') and $\xi_{\perp} = 29$ Å (//c^*) for the β_L-salt, and $\xi_{//} = 130$ Å and $\xi_{\perp} = 10$ Å for the β_H-salt.

The α-(ET)$_2$I$_3$ exhibited nearly temperature independent resistivity down to 135 K, at which charge-ordered MI transition occurred [122,123]. It has been claimed that α-(ET)$_2$I$_3$ has a zero-gap state with a Dirac cone type energy dispersion like graphene [124,125]. Under hydrostatic pressure it became two-dimensional metal down to low temperatures (2 GPa), however, it became superconductor under the uniaxial pressure along the a-axis (0.2 GPa, on-set $T_c = 7.2$ K), though along the b-axis it remained metal down to low temperature (0.3– 0.5 GPa) [117]. α-(ET)$_2$I$_3$ was able to be converted to mosaic polycrystal with $T_c \sim 8$ K by tempering at 70–100 °C for more than 3 days, giving α_t-salt which exhibited similar NMR pattern to that of the β_H-salt [36]. Other α-type superconductors (α-(ET)$_2$MHg(SCN)$_4$: M = K, NH$_4$, Rb, Tl) seem to have charge-ordered state near or next to the superconducting state with low T_c (highest $T_c = 1.7$ K for M = NH$_4$). Uniaxial strain increased T_c anisotropically (6 K at 0.5 GPa //c, 4.5 K at 1 GPa //b^* for M = NH$_4$) [126]. The salts α-(ET)$_2$MHg(SCN)$_4$ were confirmed to retain their donor packing patterns under pressure at low temperature by the SdH observation. However, for α-(ET)$_2$I$_3$ under pressure at low temperature, no exact information is reported for the donor packing in the superconducting state.

Only one θ-type superconductor, θ-(ET)$_2$I$_3$, is known, however, one third of the obtained crystals are superconducting and others remain metallic [127]. Several θ-type salts are arranged by their inter-column transfer interactions, and the dihedral angle between columns in a phase diagram showing superconducting θ-(ET)$_2$I$_3$ is next to both the charge ordered state of θ-(ET)$_2$MM'(SCN)$_4$ (M = Tl, Rb, Cs, M' = Zn, Co) and metallic phase of θ-(ET)$_2$Ag(CN)$_2$ [128]. It is not clear which point the non-superconducting θ-(ET)$_2$I$_3$ occupies in this phase diagram. Tempering (70 °C, 2 h) all crystals of θ-(ET)$_2$I$_3$ induced superconductivity with higher T_c (named θ_T-(ET)$_2$I$_3$: sharp drop of resistivity at 7 K and dull drop at ~5 K) [129]. The tempering changed the ethylene conformation and position of I_3 from disordered state in θ-phase to ordered state in θ_T-phase. Therefore, the phase diagram of the θ-type salts needs further parameters concerning with the structure of the salts (ethylene conformation and position or disorder of anions).

3.1.2.6. BETS superconductors

The most intriguing phenomenon among the 8 BETS (Scheme 6) superconductors (highest $T_c = 5.5$ K) is the reentrant superconductor-insulator-superconductor transition under magnetic field for (BETS)$_2$FeCl$_4$. The λ- or κ-type BETS salts formed with tetrahedral anions FeX$_4$ (X: Cl and Br) were studied in terms of the competition of magnetic ordering and superconductivity [130–136]. The λ-(BETS)$_2$FeCl$_4$ exhibited coupled AF and MI transitions at 8.3 K. For the FeCl$_4$ salt, a relaxor ferroelectric behavior in the metallic state below 70 K [133] and a firm nonlinear electrical transport associated with the negative resistance effect in the magnetic ordered state have been observed [135]. Moreover it has been found that the FeCl$_4$ salt shows the field-induced superconducting transition under a magnetic field of 18–41 T applied exactly parallel to the conducting layers [132]. The λ-(BETS)$_2$Fe$_x$Ga$_{1-x}$X$_4$ passes through a superconducting to insulating transition on cooling [135]. The κ-(BETS)$_2$FeX$_4$ (X = Cl, Br) are AF superconductors (T_N = 2.5 K, T_c = 1.1 K for X = Br: T_N = 0.45 K, T_c = 0.17 K for X = Cl) [136]. Similar phenomena, namely AF, ferromagnetic, or field-induced superconductivity, have been observed in several inorganic solids such as Chevrel phase [137] and heavy-fermion system [138]. Recently it has been reported that λ-(BETS)$_2$GaCl$_4$ (T_c = 8 K(on-set), 5.5 K(mid-point)) exhibited superconductivity in the minute size of four pairs of (BETS)$_2$GaCl$_4$ based on the STM study [139]. This salt has two-dimensional superconducting character ($\xi_{/\!/}$ = 125 Å, $\xi_\perp$ = 16 Å).

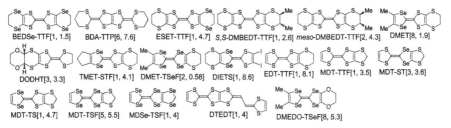

BETS

Scheme 6.

3.1.3. Superconductors of other electron donor molecules

Besides TMTSF, TMTTF, BO, ET, and BETS superconductors, there are other superconductors (Scheme 7, numbers in bracket are the total members of each superconducting family and the highest T_c) of CT salts based on symmetric (BEDSe-TTF [140] and BDA-TTP [141–144]) and asymmetric donors (ESET-TTF [145], S,S-DMBEDT-TTF [146], $meso$-DMBEDT-TTF [147,148], DMET [149], DODHT [150], TMET-STF [151], DMET-TSF [152], DIETS [153], EDT-TTF [154], MDT-TTF [155,156], MDT-ST [157,158], MDT-TS [159], MDT-TSF [160–163], MDSe-TSF [164], DTEDT [165], and DMEDO-TSeF [166,167]).

Donor Molecule

BEDSe-TTF[1, 1.5]   BDA-TTP[6, 7.6]   ESET-TTF[1, 4.7] S,S-DMBEDT-TTF[1, 2.6] *meso*-DMBEDT-TTF[2, 4.3]   DMET[8, 1.9]

DODHT[3, 3.3]   TMET-STF[1, 4.1]   DMET-TSeF[2, 0.58]   DIETS[1, 8.6]   EDT-TTF[1, 8.1]   MDT-TTF[1, 3.5]   MDT-ST[3, 3.6]

MDT-TS[1, 4.7]   MDT-TSF[5, 5.5]   MDSe-TSF[1, 4]   DTEDT[1, 4]   DMEDO-TSeF[8, 5.3]

Scheme 7. Donor molecules for organic superconductors except TMTSF, TMTTF, BO, ET, and BETS systems. Numbers in bracket are the total members of each superconductor and the highest T_c.

κ-(MDT-TTF)$_2$AuI$_2$ (T_c = 3.5 K) exhibited a Hebel-Slichter coherent peak just below T_c, indicating a BCS-type gap with s-symmetry [156]. On the other hand, d-wave like superconductivity has been suggested for β-(BDA-TTP)$_2$SbF$_6$ [143,144]. β-(BDA-TTP)$_2$X (X = SbF$_6$, AsF$_6$) exhibited a slight T_c increase at the initial stage of uniaxial strain parallel to the donor stack and interlayer direction while T_c decreased perpendicular to the donor stack [142]. θ-(DIETS)$_2$[Au(CN)$_4$] exhibited superconductivity under uniaxial strain parallel to the c-axis (T_c = 8.6 K at 1 GPa), though under hydrostatic pressure a sharp MI transition remained even at 1.8 GPa [153]. MDT-ST, MDT-TS, and MDT-TSF superconductors [157–163] have non-integer ratio of donor and anion molecules such as (MDT-TS)(AuI$_2$)$_{0.441}$ making the Fermi level different from the conventional 3/4 filled band for TMTSF and ET 2:1 salts. The Fermi surface topology of (MDT-TSF)X (X = (AuI$_2$)$_{0.436}$, (I$_3$)$_{0.422}$) and (MDT-ST)(I$_3$)$_{0.417}$ has been studied by SdH and AMRO [158,161–163]. DMEDO-TSeF afforded eight superconductors. Six of them are κ-(DMEDO-TSeF)$_2$[Au(CN)$_2$](solvent) and the T_c's of them (1.7–5.3 K) are tuned by the use of cyclic ethers as solvent of crystallization [167]. The superconducting coherent lengths indicate that β-(BDA-TTP)$_2$SbF$_6$ has two-dimensional character ($\xi_{//}$ = 105 Å, $\xi_{\perp}$ = 26 Å) while (DMET-TSeF)$_2$AuI$_2$ has quasi-one-dimensional character (1000, 400, and 20 Å).

3.2. Superconductors based on electron acceptors

Icosahedral C$_{60}$ molecule with I_h symmetry has triply degenerate LUMO and LUMO+1 orbitals with t_{1u} and t_{1g} symmetries, respectively. In 1991, superconducting phase was observed below 19 K for the potassium-doped compounds prepared by a vapor-solid reaction [168], immediately after the isolation of macroscopic quantities of C$_{60}$ solid [169]. Powder X-ray diffraction profile revealed that the composition of the superconducting phase is K$_3$C$_{60}$ and the diffraction pattern can be indexed to be a face-centered cubic (fcc) structure [170]. The lattice constant (a = 14.24 Å) is apparently expanded relative to the undoped cubic C$_{60}$ (a = 14.17 Å). The superconductivity has been observed for many M$_3$C$_{60}$ (M: alkali metal), $e.g.$, Rb$_3$C$_{60}$ (T_c = 29 K [171]), Rb$_2$CsC$_{60}$ (T_c = 31 K [172]), and RbCs$_2$C$_{60}$ (T_c = 33 K [172]), and their structures determined to be analogous to that of K$_3$C$_{60}$ with varying lattice constants. The T_c varies monotonously with lattice constant, independently of the type of the alkali dopant (Fig. 10) [172,173]. The observation of a Hebel-Slichter peak in relaxation rate just below T_c in NMR [174] and μSR [175] indicate the BCS-type isotopic gap. The decrease in T_c due to the isotopic substitution [176] also supports the phonon-mediated pairing in M$_3$C$_{60}$, where the α value in $T_c \propto$ (mass)$^{-\alpha}$ (ideal value of α predicted by the BCS model is 0.5) is estimated to be 0.30(6) for K$_3{}^{13}$C$_{60}$ and 0.30(5) for Rb$_3{}^{13}$C$_{60}$.

Keeping the C$_{60}$ valence invariant (–3), the intercalation of NH$_3$ molecules ($e.g.$, (NH$_3$)K$_3$C$_{60}$) results in the lattice distortion from cubic to orthorhombic accompanied by the appearance of AF ordering instead of superconductivity [177]. Changing the valence in cubic system also has a pronounced effect on T_c. For example, T_c in Rb$_{3-x}$Cs$_x$C$_{60}$ prepared in liquid ammonia gradually increases as the mixing ratio approaches to x = 2 [178]. Further increasing the nominal ratio of Cs leads to a sizable decrease of T_c, despite the lattice keeps the fcc structure for $x < 2.65$. Such a band-filling control has been realized for $Na_2Cs_xC_{60}$ ($0 \le x \le 1$) [179] and Li_xCsC_{60} ($2 \le x \le 6$) [180], and shows that the T_c decreases sharply as the valence state on C_{60} deviates from -3.

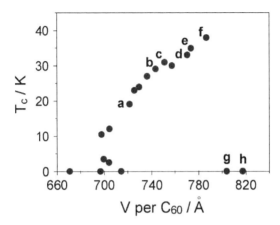

Figure 10. T_c as a function of volume occupied per C_{60}^{3-} in cubic M_3C_{60} (M: alkali metal). **a**: K_3C_{60}, **b**: Rb_3C_{60}, **c**: Rb_2CsC_{60}, **d**: $RbCs_2C_{60}$, **e**: fcc Cs_3C_{60} at 0.7 GPa, **f**: A15 Cs_3C_{60} at 0.7 GPa, **g**: fcc Cs_3C_{60} at AP, **h**: A15 Cs_3C_{60} at AP.

In 2008, A15 or body-centered cubic (bcc) Cs_3C_{60} phase, which shows the bulk superconductivity under an applied hydrostatic pressure, was obtained together with a small amount of by-products of body-centered orthorhombic (bco) and fcc phases, by a solution process in liquid methylamine [181]. Interestingly, the lattice contraction with respect to pressure results in the increase in T_c up to around 0.7 GPa, above which T_c gradually decreases. The trend in the initial pressure range is not explicable within the simple BCS theory. At AP, on the other hand, the A15 Cs_3C_{60} shows an AF ordering below 46 K, verified by means of ^{133}Cs NMR and μSR [182]. Very recently, it has been found that the fcc phase also shows an AF ordering at 2.2 K at AP and superconducting transition at 35 K under an applied hydrostatic pressure of about 0.7 GPa [183]. We note that T_c of the both phases follows the universal relationship for M_3C_{60} superconductors in the vicinity of the Mott boundary, as seen in Fig. 10. So far about 40 superconductors were prepared with the highest T_c of 33 K ($RbCs_2C_{60}$) at AP and 38 K (A15 Cs_3C_{60}) under pressure (0.7 GPa).

4. Polyaromatic hydrocarbon superconductors

Doping of alkali metals in picene (Scheme 8), that has a wider band gap (3.3 eV) than 1.8 eV for pentacene, introduced superconductivity [184]. The bulk superconducting phase was observed below 7 K and 18 K for K_3picene, and 7 K for Rb_3picene, which are comparable to that of K_3C_{60} (T_c = 19 K). Since the energy difference between LUMO and LUMO+1 is very small (< 0.1 eV), the three electrons reside in the almost two-fold degenerate LUMO.

Considering the fact that $Ca_{1.5}$picene also shows a superconducting phase below 7 K [185], three-fold charge transfer from dopants to one picene molecule would be responsible for emergence of the superconductivity. At present, although the crystal structures of the doped compounds are unclear, the refined lattice parameters are indicative of the deformation of the herringbone structure of pristine picene and the intercalation of dopants within the two-dimensional picene layers. After the discovery of the picene-based superconductors, several superconductors have been found for alkali metal (T_c = 7 K) [186] and alkaline-earth metal (T_c ~ 5.5 K) [187] doped phenanthrene, potassium-doped 1,2:8,9-dibenzopentacene (T_c = 33 K; partially decomposed) [188], and potassium-doped coronene (T_c < 15 K) [185]. Among them, phenanthrene-based superconductors shows an enhancement of T_c with increasing pressure, which is indicative of the non-BCS behavior.

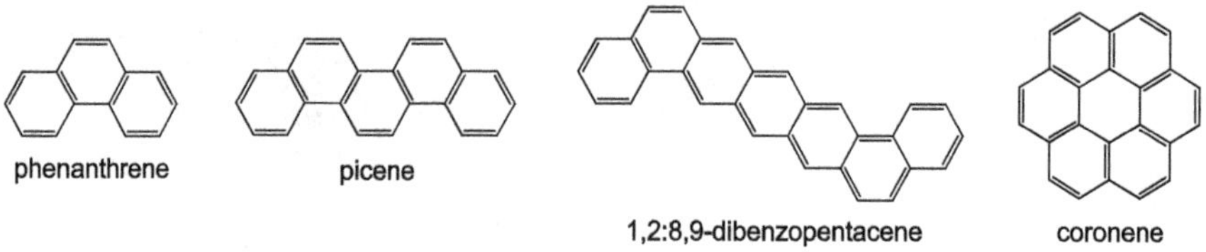

phenanthrene   picene

1,2:8,9-dibenzopentacene   coronene

Scheme 8.

Contrary to the electron-doped system above described, it has been found that a cation radical salt (perylene)$_2$Au(mnt)$_2$, in which each perylene has an average charge of +0.5 and form segregated columns, show a superconductivity with T_c = 0.3 K when the hydrostatic pressure above 0.5 GPa was applied to suppress CDW phase [189]. So far 8 aromatic hydrocarbon superconductors were prepared with the highest T_c of 33 K at AP.

5. Carbon nanotubes, graphite, and diamond superconductors

In 2001, first superconducting carbon nanotubes were discovered for ropes of single-walled carbon nanotubes (SWNTs) with diameters of the order of 1.4 nm (T_c = 0.4 K) [190], and immediately after that SWNT with diameters of 0.4 nm embedded in a zeolite matrix (T_c = 15 K) [191,192]. The drop in magnetic susceptibility is more gradual than expected for three-dimensional superconductors, and superconducting gap estimated from the I-V plot shows the temperature dependency characteristic of one-dimensional fluctuations. It is apparent that the isolation of carbon nanotubes from each other is responsible for the realization of the almost ideal one-dimensional system. Multi-walled carbon nanotubes (MWNTs) also show the superconductivity; namely, MWNT with diameters of 10–17 nm that were grown in nanopores of alumina templates was found to show superconductivity with T_c = 12 K [193]. We note that this superconducting system is classified into single component superconductors, contrary to the C_{60} and graphite (*vide infra*) based superconductors.

Graphite has a layered structure composed of infinite benzene-fused π-planes (graphenes) with sp^2 character. First-stage alkali metal doped graphite intercalation compounds (GICs) were known to superconduct with T_c = 0.15 K for KC_8 [194]. In 1980s and 1990s, further efforts were poured to synthesize GICs with higher T_c, such as LiC_2 with T_c = 1.9 K [195]. In 2005, these efforts culminated in the discovery of CaC_6 with T_c as high as 11.5 K at AP [196], which goes up to 15.1 K under pressures up to 7.5 GPa [197]. In other alkaline-earth metal doped GICs, the superconducting phase was observed below 1.65 K for SrC_6 and 6.5 K for YbC_6 [198]. The apparent reduction of T_c strongly suggests that the interlayer states of graphite have an impact on the electronic state of GIC, which was supported by theoretical calculations [199]. The conventional phonon mechanism in the framework of conventional BCS theory is generally accepted, due mainly to the observation of the Ca isotope effect with its exponent α = 0.5 [200].

A typical sp^3 covalent system, diamond, is an electrical insulator with a wide band gap of 5.5 eV, and is well known for its hardness as well as its unique electronic and thermal properties. Superconductivity in diamond was achieved through heavy p-type doping by boron in 2004 (T_c = 4 K), which was performed under high pressure (8–9 GPa) and high temperatures (2500–2800 K) [201]. Enhanced T_c in homoepitaxial CVD films has been achieved as high as 11 K [202]. Doped boron introduces an acceptor level with a hole binding energy of 0.37 eV and results in a metallic state above a critical boron concentration in the range of a few atoms per thousand. The T_c varies between 1 and 10 K with the doping level [203]. Superconducting gap estimated from STM [204] and isotopic substitution of boron and carbon [205] follow the BCS picture, as MgB_2 (T_c = 39 K).

Accordingly, all of the carbon polymorphs, namely zero-dimensional C_{60} (sp^2/sp^3 character), one-dimensional carbon nanotube (sp^2 character), two-dimensional graphite (sp^2 character), and three-dimensional diamond (sp^3 character) could provide superconductors, despite their covalent character being different. The superconductors with sp^2/sp^3 or sp^2 carbons were realized either in themselves or by doping of metal atoms, while those with sp^3 carbons were realized by substitution of boron for carbon.

6. Single component superconductors

Since the discovery of the first metallic CT solid, TTF•TCNQ, in 1973 [8], much attention for organic (super)conductors has been devoted to plural component CT solids. Besides numerous studies on multi-component CT solids, several single-component organic conductors have been developed. Even though pentacene is known to be the first organic metal (semimetal) showing a decrease of resistivity down to ca. 200 K at 21.3 GPa [206], no superconductivity was reported so far on the solids composed of aromatic hydrocarbon solely. Electric conductivity increases by the enhancement of intermolecular interactions by appropriate use of hetero-atomic contacts. There are two single-component superconductors under extremely high pressure, p-iodanil (σ_{RT} = 1 × 10^{-12} S cm^{-1} at AP, σ_{RT} = 2 × 10 S cm^{-1} at 25 GPa, and superconductor at T_c ~ 2 K at 52 GPa) [207,208] and hexaiodobenzene (T_c = 0.6–0.7 K at around 33 GPa and ca. 2.3 K at 58 GPa) [209]. Both have peripheral chalcogen atoms, iodine, which may cause the increased electronic dimensionality of the solid under pressure owing to intermolecular iodine⋯iodine contacts.

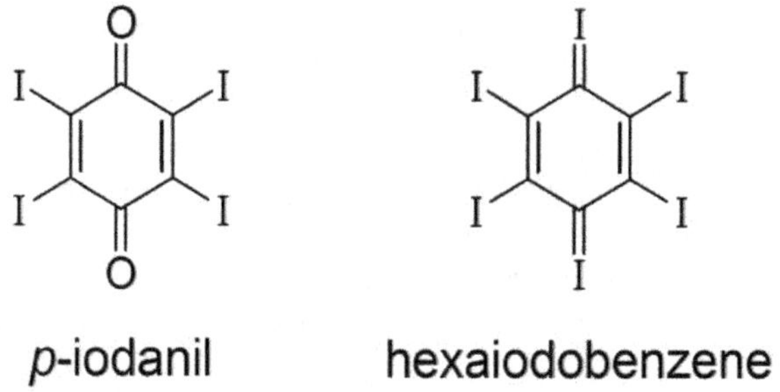

p-iodanil   **hexaiodobenzene**

Scheme 9.

Abbreviations

AF: antiferromagnetic
AMRO: angular dependent magnetoresistance oscillation
AP: ambient pressure
BCS: Bardeen-Cooper-Schrieffer
CDW: charge density wave
CT: charge transfer
$D(\varepsilon_F)$: density of states at Fermi level (ε_F) per spin
dHvA: de Haas-van Alphen
FISDW: field induced spin-density-wave
GIC: graphite intercalation compound
H_{c2}: upper critical magnetic field
H_{Pauli}: Pauli limited magnetic field
LUMO: lowest unoccupied molecular orbital
MI: metal-insulator
MWNT: multi-walled carbon nanotubes
OD: order-disorder
P_c: critical pressure for superconductivity
SdH: Shubnikov-de Haas
SDW: spin density wave
STM: scanning tunneling microscope
SWNT: single-walled carbon nanotubes
t: transfer integral
T_c: critical temperature for superconductivity
T_N: Néel temperature (temperature for AF order)
U (U_{eff}): on-site Coulomb repulsion energy (effective U)
W: band width
Θ: Debye temperature

Author details

Gunzi Saito and Yukihiro Yoshida

Faculty of Agriculture, Meijo University, Shiogamaguchi 1-501 Tempaku-ku, Nagoya, Japan

7. References

[1]   Little W A (1964) Phys. Rev. A134: 1416-1424.

[2]   Greene R L, Street G B, Suter L J (1975) Phys. Rev. Lett. 34: 577-579.

[3]   Kawamura H, Shirotani I, Tachikawa K (1984) Solid State Commun. 49: 879-881.

[4]   Lee K, Cho S, Park S H, Heeger A J, Lee C-W, Lee S-H (2006) Nature 441: 65-68.

[5]   As a review see for example, Prassides K (2000) In: Andreoni W, editor. The Physics of Fullerene-Based and Fullerene-Related Materials. Boston: Kluwer Academic Publishers. pp. 175-202.

[6]   Akamatu H, Inokuchi H, Matsunaga Y (1954) Nature 173: 168-169.

[7]   As a review see for example, Hertler W R, Mahler W, Melby L R, Miller J S, Putscher R E, Webster O W (1989) Mol. Cryst. Liq. Cryst. 171: 205-216.

[8]   Ferraris J, Cowan D O, Walatka V, Perlstein J H (1973) J. Am. Chem. Soc. 95: 948-949.

[9]   Jerome D, Mazaud A, Ribault M, Bechgaard K (1980) J. Phys. Lett. 41: L95-L98.

[10] Lee I J, Naughton M J, Tanner G M, Chaikin P M (1997) Phys. Rev. Lett. 78: 3555-3558.

[11] Wudl F, Aharon-Shalom E, Nalewajek D, Waszeczak J V, Walsh Jr. W M, Rupp Jr. L W, Chaikin P M, Lacoe R, Burns M, Poehler T O, Williams J M, Beno M A, J. Chem. Phys. (1982) 76: 5497-5501.

[12] Sakata M, Yoshida Y, Maesato M, Saito G, Matsumoto K, Hagiwara R (2006) Mol. Cryst. Liq. Cryst. 452: 99-108.

[13] Jerome D, Schulz H J (1982) Adv. Phys. 31: 299-490.

[14] Jerome D (2004) Chem. Rev. 104: 5565-5591.

[15] Horiuchi S, Yamochi H, Saito G, Sakaguchi K, Kusunoki M (1996) J. Am. Chem. Soc. 118: 8604-8622.

[16] Saito G, Yoshida Y (2007) Bull. Chem. Soc. Jpn. 80: 1-137 and references cited therein.

[17] Ishiguro T, Yamaji K, Saito G (1998) Organic Superconductors, 2nd ed.: Springer-Verlag, Berlin.

[18] Williams J M, Ferraro J R, Thorn R J, Carlson K D, Geiser U, Wang H H, Kini A M, Whangbo M -H (1992) Organic Superconductors (Including Fullerenes): Prentice Hall, Englewood Cliffs, NJ.

[19] Wosnitza J (1996) Fermi Surfaces of Low-Dimensional Organic Metals and Superconductors: Springer-Verlag, Berlin.

[20] Singleton J (2000) Rep. Prog. Phys. 63: 1111-1207.

[21] Chem. Rev. (2004) 104, No. 11, Special Issue on Molecular Conductors.

[22] Balicas L, Behnia K, Kang W, Canadell E, Auban-Senzier P, Jerome D, Ribault M, Fabre J M (1994) J. Phys. I (France) 4: 1539-1549.

[23] Adachi T, Ojima E, Kato K, Kobayashi H, Miyazaki T, Tokumoto M, Kobayashi A (2000) J. Am. Chem. Soc. 122: 3238-3239.

[24] Auban-Senzier P, Pasquier C, Jerome D, Carcel C, Fabre J M (2003) Synth. Met. 133-134: 11-14.

[25] Itoi M, Araki C, Hedo M, Uwatoko Y, Nakamura T (2008) J. Phys. Soc. Jpn. 77: 023701/1-4.

[26] Takigawa M, Yasuoka H, Saito G (1987) J. Phys. Soc. Jpn. 56: 873-876.

[27] Hasegawa Y, Fukuyama H (1987) J. Phys. Soc. Jpn. 56: 877-880.

[28] Belin S, Behnia K (1997) Phys. Rev. Lett. 79: 2125-2128.

[29] Jerome D (1991) Science 252: 1509-1514. A more detail diagram was then proposed; Dumm M, Loidl A, Fravel B W, Starkey K P, Montgomery L K, Dressel M (2000) Phys. Rev. B61: 511-521.

[30] Beno M A, Wang H H, Kini A M, Carlson K D, Geiser U, Kwok W K, Thompson J E, Williams J M, Ren J, Whangbo M -H (1990) Inorg. Chem. 29: 1599-1601.

[31] Kahlich S, Schweitzer D, Heinen I, Lan S E, Nuber B, Keller H J, Winzer K, Helberg H W (1991) Solid State Commun. 80: 191-195.

[32] Saito G, Enoki T, Toriumi K, Inokuchi H (1982) Solid State Commun. 42: 557-560.

[33] Leung P C W, Emge T J, Beno M A, Wang H H, Williams J M, Patrick V, Coppens P (1985) J. Am. Chem. Soc. 107: 6184-6191.

[34] Saito G, Otsuka A, Zakhidov A A (1996) Mol. Cryst. Liq. Cryst. 284: 3-14.

[35] Baram O, Buravov L I, Degtyarev L S, Kozlov M E, Laukhin V N, Laukhina E E, Onishchenko V G, Pokhodnya K I, Sheinkman M K, Shibaeva R P, Yagubskii E B (1986) JETP Lett. 44: 376-378.

[36] Schweitzer D, Bele P, Brunner H, Gogu E, Haeberlen U, Hennig I, Klutz I, Sweitlik R, Keller H J (1987) Z. Phys. B 67: 489-495.

[37] Yoshimura M, Shigekawa H, Yamochi H, Saito G, Kawazu A (1991) Phys. Rev. B44: 1970-1972.

[38] Yoshimura M, Shigekawa H, Nejoh H, Saito G, Saito Y, Kawazu A (1991) Phys. Rev. B43: 13590-13593.

[39] Yoshimura M, Ara N, Kageshima M, Shiota R, Kawazu A, Shigekawa H, Saito Y, Oshima M, Mori H, Yamochi H, Saito G (1991) Surf. Sci. 242: 18-22.

[40] Yamochi H, Komatsu T, Matsukawa N, Saito G, Mori T, Kusunoki K, Sakaguchi K (1993) J. Am. Chem. Soc. 115: 11319-11327.

[41] Kobayashi A, Kato R, Kobayashi H, Moriyama S, Nishio Y, Kajita K, Sasaki W (1987) Chem. Lett. 459-462.

[42] Yagubskii E B, Shchegolev I F, Laukhin V N, Kononovich P A, Karstovnik M V, Zvarykina A V, Buravov L I (1984) JETP Lett. 39: 12-15.

[43] Shibaeva R P, Kaminskii V F, Yagubskii E B (1985) Mol. Cryst. Liq. Cryst. 119: 361-373.

[44] Kagoshima S, Mori H, Nogami Y, Kinoshita N, Anzai H, Tokumoto M, Saito G (1990) In: Saito G, Kagoshima S, editors. The Physics and Chemistry of Organic Superconductors: Springer-Verlag, Berlin, pp. 126-129.

[45] Laukhin V N, Kostyuchenko E E, Sushko Yu V, Shchegolev I F, Yagubskii E B (1985) Pis'ma Zh. Eksp. Teor. Fiz. 41: 68-70.

[46] Murata K, Tokumoto M, Anzai H, Bando H, Saito G, Kajimura K, Ishiguro T (1985) J. Phys. Soc. Jpn. 54: 1236-1239.

[47] Schirber J E, Azevedo L J, Kwak J F, Venturini E L, Leung P C W, Beno M A, Wang H H, Williams J M (1986) Phys. Rev. B33: 1987-1989.

[48] Wang H H, Nunez L, Carlson G W, Williams J M, Azevedo J L, Kwak J F, Schirber J E (1985) Inorg. Chem. 24: 2465-2466.

[49] Komatsu T, Nakamura T, Matsukawa N, Yamochi H, Saito G (1991) Solid State Commun. 80: 843-847.

[50] Fig.54 in Ref. 16.

[51] Urayama H, Yamochi H, Saito G, Nozawa K, Sugano T, Kinoshita M, Sato S, Oshima K, Kawamoto A, Tanaka J (1988) Chem. Lett. 55-58.

[52] Urayama H, Yamochi H, Saito G, Sato S, Kawamoto A, Tanaka J, Mori T, Maruyama Y, Inokuchi H (1988) Chem. Lett. 463-466.

[53] Oshima K, Mori T, Inokuchi H, Urayama H, Yamochi H, Saito G (1988) Phys. Rev. B38: 938-941.

[54] Fig. 53 in Ref. 16.

[55] Kini M, Geiser U, Wang H H, Carlson K D, Williams J M, Kwok W K, Vandervoort K G, Thompson J E, Stupka D L, Jung D, Wangbo M -H (1990) Inorg. Chem. 29: 2555-2557.

[56] Komatsu T, Matsukawa N, Nakamura T, Yamochi H, Saito G (1992) Phosphorus, Sulfur Silicon Relat. Elem. 67: 295-300, and Section 3-5-4-2 in Ref. 16.

[57] Williams J M, Kini A M, Wang H H, Carlson K D, Geiser U, Montgomery L K, Pyrka G J, Watkins D M, Kommers J M, Boryschuk S J, Crouch A V, Kwok W K, Schirber J E, Overmyer D L, Jung D, Whangbo M -H (1990) Inorg. Chem. 29: 3272-3274.

[58] Schirber J E, Overmyer D L, Carlson K D, Williams J M, Kini A M, Wang H H, Charlier H A, Love B J, Watkins D M, Yaconi G A (1991) Phys. Rev. B44: 4666-4669.

[59] Welp U, Fleshler S, Kwok W K, Crabtree G W, Carlson K D, Wang H H, Geiser U, Williams J M, Hitsman V M (1992) Phys. Rev. Lett. 69: 840-843. This paper mistakenly described that the AF transition occurred at 45 K and weak ferromagnetic transition at 23 K. The magnetic susceptibility measurements could not identify the transition either AF or SDW.

[60] Miyagawa K, Kawamoto K, Nakazawa Y, Kanoda K (1995) Phys. Rev. Lett. 75: 1174-1177.

[61] Geiser U, Wang H H, Carlson K D, Williams J M, Charlier Jr. H A, Heindl J E, Yaconi G A, Love B H, Lathrop M W, Schirber J E, Overmyer D L, Ren J, Whangbo M -H (1991) Inorg. Chem. 30: 2586-2588.

[62] Shimizu Y, Maesato M, Saito G, Drozdova O, Ouahab L (2003) Synth. Met. 133-134: 225-226.

[63] Shimizu Y, Miyagawa K, Kanoda K, Maesato M, Saito G (2003) Phys. Rev. Lett. 91: 107001/1-4.

[64] Pratt F L, Baker P J, Blundell S J, Lancaster T, Ohira-Kawamura S, Baines C, Shimizu Y, Kanoda K, Watanabe I, Saito G (2011) Nature 471: 612-616.

[65] Shimizu Y, Maesato M, Saito G (2011) J. Phys. Soc. Jpn. 80: 074702/1-7.

[66] Iwasa Y, Mizuhashi K, Koda T, Tokura Y, Saito G (1994) Phys. Rev. B49: 3580-3583.

[67] Yoneyama N, Miyazaki A, Enoki T, Saito G (1999) Bull. Chem. Soc. Jpn. 72: 639-651.

[68] Taniguchi H, Miyashita M, Uchiyama K, Satoh K, Mori N, Okamoto H, Miyagawa K, Kanoda K, Hedo M, Uwatoko Y (2003) J. Phys. Soc. Jpn. 72: 468-471.

[69] Uchiyama K, Miyashita M, Taniguchi H, Satoh K, Mori N, Miyagawa K, Kanoda K, Hedo M, Uwatoko Y (2004) J. Phys. IV 114: 387-389.

[70] Appendix (pp. 455-458) in Ref. 17.

[71] Sasaki T, Sato H, Toyota N (1990) Solid State Commun. 76: 507-510.

[72] Ito H, Watanabe M, Nogami Y, Ishiguro T, Komatsu T, Saito G, Hosoito N (1991) J. Phys. Soc. Jpn. 60: 3230-3233.

[73] Sushko Y V, Ito H, Ishiguro T, Horiuchi S, Saito G (1993) J. Phys. Soc. Jpn. 62: 3372-3375.

[74] Ito H, Ishiguro T, Kubota M, Saito G (1996) J. Phys. Soc. Jpn. 65: 2987-2993.

[75] Kubota M, Saito G, Ito H, Ishiguro T, Kojima N (1996) Mol. Cryst. Liq. Cryst. 284: 367-377.

[76] Ito H, Ishiguro T, Kondo T, Saito G (2000) J. Phys. Soc. Jpn. 69: 290-291.

[77] Posselt H, Muller H, Andres K, Saito G (1994) Phys. Rev. B49: 15849-15852.

[78] Lefebvre S, Wzietek P, Brown S, Bourbonnais C, Jérome D, Mézière C, Fourmigué M, Batial P (2000) Phys. Rev. Lett. 85: 5420-5423.

[79] Tanatar M A, Ishiguro T, Kagoshima S, Kushch N D, Yagubskii E B (2002) Phys. Rev. B65: 064516/1-5.

[80] Kanoda K (2006) J. Phys. Soc. Jpn. 75: 051007/1-16.

[81] Schulueter J A, Wiehl L, Park H, de Souza M, Lang M, Koo H -J, Whangbo M -H (2010) J. Am. Chem. Soc. 132: 16308-16310.

[82] Kawamoto T, Mori T, Nakao A, Murakami Y, Schlueter J A (2012) J. Phys. Soc. Jpn. 81: 023705/1-4.

[83] Pereiro J, Petrovic A, Panagopoulos C, Božović I (2011) Phys. Express 1: 208-241.

[84] Gariglio S, Triscone J -M (2011) C. R. Phys. 12: 591-599.

[85] Kartsovnik M V, Logvenov G Yu, Ito H, Ishiguro T, Saito G (1995) Phys. Rev. B52: R15715- R15718.

[86] Weiss H, Kartsovnik M V, Biberacher W, Steep E, Balthes E, Jansen A G M, Andres K, Kushch N D (1999) Phys. Rev. B59: 12370-12378.

[87] Kartsovnik M V, Biberacher W, Andres K, Kushch N D (1995) JETP Lett. 62: 905-909.

[88] Oshima K, Urayama H, Yamochi H, Saito G (1988) J. Phys. Soc. Jpn. 57: 730-733.

[89] Nam M -S, Symington J A, Singleton J, Blundell S J, Ardavan A, Perenboom J A A J, Kurmoo M, Day P (1999) J. Phys.: Condens. Matter 11: L477-484.

[90] Arai T, Ichimura K, Nomura K (2001) Phys. Rev. B63: 104518/1-5.

[91] Izawa K, Yamaguchi H, Sasaki T, Matsuda Y (2002) Phys. Rev. Lett. 88: 27002/1-4.

[92] Ichimura K, Takami M, Nomura K (2008) J. Phys. Soc. Jpn. 77: 114707/1-6.

[93] Malone L, Taylor O J, Schlueter J A, Carrington A (2010) Phys. Rev. B82: 014522/1-5.

[94] Toyota N (1996) In: Physical Phenomena at High Magnetic Fields-II. World Sci., pp. 282-293.

[95] McKenzie R H (1998) Comments Condens. Matter Phys. 18: 309-337.

[96] Kino H, Fukuyama H (1995) J. Phys. Soc. Jpn. 64: 2726-2729.

[97] Wannier G H (1950) Phys. Rev. 79: 357-364.

[98] Anderson P W (1973) Mater. Res. Bull. 8: 153-164.

[99] Balents L (2010) Nature 464: 199-208.

[100] Norman M R (2011) Science 332: 196-200.

[101] Furukawa Y, Sumida Y, Kumagai K, Borsa F, Nojiri H, Shimizu Y, Amitsuka H, Tenya K, Kögerler P, Cronin L, (2011) J. Phys.: Conf. Ser. 320: 012047/1-6.

[102] Seeber G, Kögerler P, Kariuki B M, Cronin L (2004) Chem. Commun. 1580-1581.

[103] Itou T, Oyamada A, Maegawa S, Tamura M, Kato R (2007) J. Phys.: Condens. Matter 19: 145247/1-5.

[104] Shores M P, Nytko E A, Bartlett B M, Nocera D G (2005) J. Am. Chem. Soc. 127: 13462-13463.

[105] Mendels P, Bert F (2010) J. Phys. Soc. Jpn. 79: 011001/1-10.

[106] Okamoto Y, Nohara M, Aruga-Katori H, Takagi H (2007) Phys. Rev. Lett. 99: 137207/1-4.

[107] Okamoto Y, Yoshida H, Hiroi Z (2009) J. Phys. Soc. Jpn. 78: 033701/1-4.

[108] Colman R H, Bert F, Boldrin D, Hiller A D, Manuel P, Mendels P, Wills A S (2011) Phys. Rev. B83: 180416/1-4.

[109] Coldea R, Tennante D A, Tylczynski Z (2003) Phys. Rev. B68: 134424/1-16.

[110] Nakatsuji S, Nambu Y, Tonomura H, Sakai O, Broholm C, Tsunetsugu H, Qiu Y, Maeno Y (2005) Science 309: 1697-1700.

[111] Olariu A, Mendels P, Bert F, Ueland B G, Schiffer P, Berger R F, Cava R J (2006) Phys. Rev. Lett. 97: 167203/1-4.

[112] Kimmel A, Mucksch M, Tsurkan V, Koza M M, Mutka H, Loidl A (2005) Phys. Rev. Lett. 94: 237402/1-4.

[113] Yoshida H, Okamoto Y, Tayama T, Sakakibara T, Tikunaga M, Matsuo A, Narumi Y, Kindo K, Yoshida M, Takigawa M, Hiroi Z (2009) J. Phys. Soc. Jpn. 78: 043704/1-4.

[114] Iwasa Y, Takenobu T (2003) J. Phys.: Condens. Matter 15: R495-R519.

[115] Chu C W (2009) Nat. Phys. 5: 787-789.

[116] Maesato M, Shimizu Y, Ishikawa T, Saito G, Miyagawa K, Kanoda K (2004) J. Phys. IV 114: 227-232.

[117] Tajima N, Ebina-Tajima A, Tamura M, Nishio Y, Kajita K (2002) J. Phys. Soc. Jpn. 71: 1832-1835.

[118] Shibaeva R P, Kaminskii V F, Yagubskii E B (1985) Mol. Cryst. Liq. Cryst. 119: 361-373.

[119] Kobayashi H, Kato R, Kobayashi A, Nishio Y, Kajita K, Sasaki W (1986) Chem. Lett. 789-792.

[120] Ito H, Ishihara T, Niwa M, Suzuki T, Onari S, Tanaka Y, Yamada J, Yamochi H, Saito G (2010) Physica B 405: S262-264.

[121] Emge T J, Leung P C W, Beno M A, Schultz A J, Wang H H, Sowa L M, Williams J M (1984) Phys. Rev. B30: 6780-6782.

[122] Bender K, Dietz K, Endres H, Helberg H V, Hennig I, Keller H J, Schafer H V, Schweitzer D (1984) Mol. Cryst. Liq. Cryst. 107: 45-53.

[123] Takano Y, Hiraki K, Yamamoto H M, Nakamura T, Takahashi T (2001) J. Phys. Chem. Solids 62: 393-395.

[124] Tajima N, Sugawara S, Tamura M, Nishio Y, Kajita K (2006) J. Phys. Soc. Jpn. 75: 051010/1-10.

[125] Katayama S, Kobayashi A, Suzumura Y (2006) J. Phys. Soc. Jpn. 75: 054705/1-6.

[126] Kondo R, Kagoshima S, Maesato M (2003) Phys. Rev. B67: 134519/1-6.

[127] Kobayashi H, Kato R, Kobayashi A, Nishio Y, Kajita K, Sasaki W (1986) 789-792: 833-836.

[128] Mori H, Tanaka S, Mori T (1998) Phys. Rev. B57: 12023-12029.

[129] Salameh B, Nothardt A, Balthes E, Schmidt W, Schweitzer D, Strempfer J, Hinrichsen B, Jansen M, Maude D K (2007) Phys. Rev. B75: 054509/1-13.

[130] Tanaka H, Ojima E, Fujiwara H, Nakazawa Y, Kobayashi H, Kobayashi A (2000) J. Mater. Chem. 10: 245-247.

[131] Kobayashi H, Tomita H, Naito T, Kobayashi A, Sakai F, Watanabe T, Cassoux P (1996) J. Am. Chem. Soc. 118: 368-377.

[132] Uji S, Shinagawa H, Terashima T, Yakabe T, Terai Y, Tokumoto M, Kobayashi A, Tanaka H, Kobayashi H (2001) Nature 410: 908-910.

[133] Matsui H, Tsuchiya H, Suzuki T, Negishi E, Toyota N (2003) Phys. Rev. B68: 155105/1-10.

[134] Toyota N, Abe Y, Matsui H, Negishi E, Ishizaki Y, Tsuchiya H, Uozaki H, Endo S (2002) Phys. Rev. B66: 033201/1-4.

[135] Kobayashi H, Sato A, Arai E, Akutsu H, Kobayashi A, Cassoux P (1997) J. Am. Chem. Soc. 119: 12392-12393.

[136] Fujiwara E, Fujiwara H, Kobayashi H, Otsuka T, Kobayashi A (2002) Adv. Mater. 14: 1376-1379.

[137] Meul H W, Rossel C, Decroux M, Fischer Ø, Remenyi G, Briggs A (1984) Phys. Rev. Lett. 53: 497-500.

[138] Lin C L, Teter J, Crow J E, Mihalisin T, Brooks J, Abou-Aly A I, Stewart G R (1985) Phys. Rev. Lett. 54: 2541-2544.

[139] Clark K, Hassanien A, Khan S, Braun K -F, Tanaka H, Hla S -W (2010) Nature Nanotech. 5: 261-265.

[140] Sakata J, Sato H, Miyazaki A, Enoki T, Okano Y, Kato R (1998) Solid State Commun. 108: 377-381.

[141] Yamada J, Watanabe M, Akutsu H, Nakatsuji S, Nishikawa H, Ikemoto I, Kikuchi K (2001) J. Am. Chem. Soc. 123: 4174-4180.

[142] Ito H, Ishihara T, Tanaka H, Kuroda S, Suzuki T, Onari S, Tanaka Y, Yamada J, Kikuchi K (2008) Phys. Rev. B78: 172506/1-4.

[143] Shimojo Y, Ishiguro T, Toita T, Yamada J (2002) J. Phys. Soc. Jpn. 71: 717-720.

[144] Nomura K, Muraoka R, Matsunaga N, Ichimura K, Yamada J (2009) Physica B 404: 562-564.

[145] Okano Y, Iso M, Kashimura Y, Yamaura J, Kato R (1999) Synth. Met. 102: 1703-1704.

[146] Zambounis J S, Mayer C W, Hauenstein K, Hilti B, Hofherr W, Pfeiffer J, Buerkle M, Rihs G (1992) Adv. Mater. 4: 33-35.

[147] Kimura S, Maejima T, Suzuki H, Chiba R, Mori H, Kawamoto T, Mori T, Moriyama H, Nishio Y, Kajita K (2004) Chem. Commun. 2454-2455.

[148] Kimura S, Suzuki H, Maejima T, Mori H, Yamaura J, Kakiuchi T, Sawa H, Moriyama H (2006) J. Am. Chem. Soc. 128: 1456-1457.

[149] Kikuchi K, Murata K, Honda Y, Namiki T, Saito K, Ishiguro T, Kobayashi K, Ikemoto I (1987) J. Phys. Soc. Jpn. 56: 3436-3439.

[150] Nishikawa H, Morimoto T, Kodama T, Ikemoto I, Kikuchi K, Yamada J, Yoshino H, Murata K (2002) J. Am. Chem. Soc. 124: 730-731.

[151] Kato R, Yamamoto K, Okano Y, Tajima H, Sawa H (1997) Chem. Commun. 947-948.

[152] Kato R, Aonuma S, Okano Y, Sawa H, Tamura M, Kinoshita M, Oshima K, Kobayashi A, Bun K, Kobayashi H (1993) Synth. Met. 61: 199-206.

[153] Imakubo T, Tajima N, Tamura M, Kato R, Nishio Y, Kajita K (2002) J. Mater. Chem. 12: 159-161.

[154] Lyubovskaya R N, Zhilyaeva E I, Torunova S A, Mousdis G A, Papavassiliou G C, Perenboom J A A J, Pesotskii S I, Lyubovskii R B (2004) J. Phys. IV 114: 463-466.

[155] Papavassiliou G C, Mousdis G A, Zambounis J S, Terzis A, Hountas A, Hilti B, Mayer C W, Pfeiffer J, Synth. Met. (1988) B27: 379-383.

[156] Takahashi T, Kobayashi Y, Nakamura T, Kanoda K, Hilti B, Zambounis J S (1994) Physica C235-240: 2461-24562.

[157] Takimiya K, Takamori A, Aso Y, Otsubo T, Kawamoto T, Mori T (2003) Chem. Mater. 15: 1225-1227.

[158] Kawamoto T, Mori T, Enomoto K, Koike T, Terashima T, Uji S, Takamori A, Takimiya K, Otsubo T (2006) Phys. Rev. B73: 024503/1-5.

[159] Takimiya K, Kodani M, Niihara N, Aso Y, Otsubo T, Bando Y, Kawamoto T, Mori T (2004) Chem. Mater. 16: 5120-5123.

[160] Takimiya K, Kataoka Y, Aso Y, Otsubo T, Fukuoka H, Yamanaka S (2001) Angew. Chem., Int. Ed. 40: 1122-1125.

[161] Kawamoto T, Mori T, Konoike T, Enomoto K, Terashima T, Uji S, Kitagawa H, Takimiya K, Otsubo T (2006) Phys. Rev. B73: 094513/1-8.

[162] Kawamoto T, Mori T, Terakura C, Terashima T, Uji S, Takimiya K, Aso Y, Otsubo T (2003) Phys. Rev. B67: 020508/1-4.

[163] Kawamoto T, Mori T, Terakura C, Terashima T, Uji S, Tajima H, Takimiya K, Aso Y, Otsubo T (2003) Eur. Phys. J. B 36: 161-167.

[164] Kodani M, Takamori A, Takimiya K, Aso Y, Otsubo T (2002) J. Solid State Chem. 168: 582-589.

[165] Misaki Y, Higuchi N, Fujiwara H, Yamabe T, Mori T, Mori H, Tanaka S (1995) Angew. Chem., Int. Ed. Engl. 34: 1222-1225.

[166] Shirahata T, Kibune M, Imakubo T (2006) Chem. Commun. 1592-1594.

[167] Shirahata T, Kibune M, Yoshino H, Imakubo T (2007) Chem. Eur. J. 13: 7619-7630.

[168] Hebard A F, Rosseinsky M J, Haddon R C, Murphy D W, Glarum S H, Palstra T T M, Ramirez A P, Kortan A R (1991) Nature 350: 600-601.

[169] Krätschmer W, Lamb L D, Fostiropoulos K, Huffman D R (1990) Nature 347: 354-358.

[170] Stephens P W, Mihaly L, Lee P L, Whetten R L, Huang S -M, Kaner R, Diederich F, Holczer K (1991) Nature 351: 632-634.

[171] Rosseinsky M J, Ramirez A P, Glarum S H, Murphy D W, Haddon R C, Hebard A F, Palstra T T M, Kortan A R, Zahurak S M, Makhija A V (1991) Phys. Rev. Lett. 66: 2830-2832.

[172] Tanigaki K, Ebbesen T W, Saito S, Mizuki J, Tsai J -S, Kubo Y, Kuroshima S (1991) Nature 352: 222-223.

[173] Fleming R M, Ramirez A P, Rosseinsky M J, Murphy D W, Haddon R C, Zahurak S M, Makhija A V (1991) Nature 352: 787-788.

[174] Sasaki S, Matsuda A, Chu C W (1994) J. Phys. Soc. Jpn. 63: 1670-1673.

[175] Kiefl R F, MacFarlane W A, Chow K H, Dunsiger S, Duty T L, Johnston T M S, Schneider J W, Sonier J, Brard L, Strongin R M, Fischer J E, Smith III A B (1993) Phys. Rev. Lett. 70: 3987-3990.

[176] Chen C -C, Lieber C M (1993) Science 259: 655-658.

[177] Takenobu T, Muro T, Iwasa Y, Mitani T (2000) Phys. Rev. Lett. 85: 381-384.

[178] Dahlke P, Denning M S, Henry P F, Rosseinsky M J (2000) J. Am. Chem. Soc. 122: 12352-12361.

[179] Yildirim T, Barbedette L, Fischer J E, Lin C L, Robert J, Petit P, Palstra T T M (1996) Phys. Rev. Lett. 77: 167-170.

[180] Kosaka M, Tanigaki K, Prassides K, Margadonna S, Lappas A, Brown C M, Fitch A N (1999) Phys. Rev. B59: R6628-R6630.

[181] Ganin A Y, Takabayashi Y, Khimyak Y Z, Margadonna S, Tamai A, Rosseinsky M J, Prassides K (2008) Nat. Mat. 7: 367-371.

[182] Takabayashi Y, Ganin A Y, Jeglič P, Arčon D, Takano T, Iwasa Y, Ohishi Y, Takata M, Takeshita N, Prassides K, Rosseinsky M J (2009) Science 323: 1585-1590.

[183] Ganin A Y, Takabayashi Y, Jeglič P, Arčon D, Potočnik A, Baker P J, Ohishi Y, McDonald M T, Tzirakis M D, McLennan A, Darling G R, Takata M, Rosseinsky M J, Prassides K (2010) Nature 466: 221-225.

[184] Mitsuhashi R, Suzuki Y, Yamanari Y, Mitamura H, Kambe T, Ikeda N, Okamoto H, Fujiwara A, Yamaji M, Kawasaki N, Maniwa Y, Kubozono Y (2010) Nature 464: 76-79.

[185] Kubozono Y, Mitamura H, Lee X, He X, Yamanari Y, Takahashi Y, Suzuki Y, Kaji Y, Eguchi R, Akaike K, Kambe T, Okamoto H, Fujiwara A, Kato T, Kosugi T, Aoki H (2011) Phys. Chem. Chem. Phys. 13: 16476-16493.

[186] Wang X F, Liu R H, Gui Z, Xie Y L, Yan Y J, Ying J J, Luo X G, Chen X H (2011) Nat. Commun. 2: 507-513.

[187] Wang X F, Yan Y J, Gui Z, Liu R H, Ying J J, Luo X G, Chen X H (2011) Phys. Rev. B84: 214523/1-4.

[188] Xue M, Cao T, Wang D, Wu Y, Yang H, Dong X, He J, Li F, Chen G F (2012) Sci. Rep. 2: 389/1–4.

[189] Graf D, Brooks J S, Almeida M, Dias J C, Uji S, Terashima T, Kimata M (2009) Europhys. Lett. 85: 27009/1-5.

[190] Kociak M, Kasumov A Yu, Guéron S, Reulet B, Khodos I I, Gorbatov Yu B, Volkov V T, Vaccarini L, Bouchiat H (2001) Phys. Rev. Lett. 86: 2416-2419.

[191] Tang Z K, Zhang L, Wang N, Zhang X X, Wen G H, Li G D, Wang J N, Chan C T, Sheng P (2001) Science 292: 2462-2465.

[192] Lortz R, Zhang Q, Shi W, Ye J T, Qiu C, Wang Z, He H, Sheng P, Qian T, Tang Z, Wang N, Zhang X, Wang J, Chan C T (2009) Proc. Natl. Acad. Sci. USA 106: 7299-7303.

[193] Takesue I, Haruyama J, Kobayashi N, Chiashi S, Maruyama S, Sugai T, Shinohara H (2006) Phys. Rev. Lett. 96: 057001/1-4.

[194] Hannay N B, Geballe T H, Matthias B T, Andres K, Schmidt P, MacNair D (1965) Phys. Rev. Lett. 14: 225-226.

[195] Belash I T, Bronnikov A D, Zharikov O V, Pal'nichenko A V (1989) Solid State Commun. 69: 921-923.

[196] Weller T E, Ellerby M, Saxena S S, Smith R P, Skipper N T (2005) Nat. Phys. 1: 39-41.

[197] Gauzzi A, Takashima S, Takeshita N, Terakura C, Takagi H, Emery N, Hérold C, Lagrange P, Loupias, G (2007) Phys. Rev. Lett. 98: 067002/1-4.

[198] Kim J S, Boeri L, O'Brien J R, Razavi F S, Kremer R K (2007) Phys. Rev. Lett. 99: 027001/1-4.

[199] Csányi G, Littlewood P B, Nevidomskyy A H, Pikard C J, Simons B D (2005) Nat. Phys. 1: 42-45.

[200] Hinks D G, Rosenmann D, Claus H, Bailey M S, Jorgensen J D (2007) Phys. Rev. B75: 014509/1-6.

[201] Ekimov E A, Sidorov V A, Bauer E D, Mel'nik N N, Curro N J, Thompson J D, Stishov S M (2004) Nature 428: 542-545.

[202] Takano Y, Takenouchi T, Ishii S, Ueda S, Okutsu T, Sakaguchi I, Umezawa H, Kawarada H, Tachiki M (2007) Diamond Relat. Mater. 16: 911-914.

[203] Bustarret E, Kačmarčik J, Marcenat C, Gheeraert E, Cytermann C, Marcus J, Klein T (2004) Phys. Rev. Lett. 93: 237005/1-4.

[204] Sacépé B, Chapelier C, Marcenat C, Kačmarčik J, Klein T, Bernard M, Bustarret E (2006) Phys. Rev. Lett. 96: 097006/1-4.

[205] Ekimov E A, Sidorov V A, Zoteev A V, Lebed J B, Thompson J D, Stishov S M (2008) Sci. Technol. Adv. Mater. 9: 044210/1-6.

[206] Aust R B, Bentley W H, Drickamer H G (1964) J. Chem. Phys. 41: 1856-1864.

[207] Yokota T, Takeshita N, Shimizu K, Amaya K, Onodera A, Shirotani I, Endo S (1996) Czech. J. Phys. 46: 817-818.

[208] Amaya K, Shimizu K, Takeshita N, Eremets M I, Kobayashi T C, Endo S (1998) J. Phys.: Condens. Matter 10: 11179-11190.

[209] Iwasaki E, Shimizu K, Amaya K, Nakayama A, Aoki K, Carlon R P (2011) Synth. Met. 120: 1003-1004.

Chapter 6

Superconducting Magnet Technology and Applications

Qiuliang Wang, Zhipeng Ni and Chunyan Cui

Additional information is available at the end of the chapter

1. Introduction

The development of superconducting magnet science and technology is dependent on higher magnetic field strength and better field quality. The high magnetic field is an exciting cutting-edge technology full of challenges and also essential for many significant discoveries in science and technology, so it is an eternal scientific goal for scientists and engineers. Combined with power-electronic devices and related software, the entire magnet system can be built into various scientific instruments and equipment, which can be found widely applied in scientific research and industry. Magnet technology plays a more and more important role in the progress of science and technology. The ultra-high magnetic field helps us understand the world much better and it is of great significance for the research into the origins of life and disease prevention. Electromagnetic field computation and optimization of natural complex magnet structures pose many challenging problems. The design of modern magnets no longer relies on simple analytical calculations because of the complex structure and harsh requirements. High-level numerical analysis technology has been widely studied and applied in the large-scale magnet system to decide the electromagnetic structure parameters. Since different problems have different properties, such as geometrical features, the field of application, function and material properties, there is no single method to handle all possible cases. Numerical analysis of the electromagnetic field distribution with respect to space and time can be done by solving the Maxwell's equations numerically under predefined initial and boundary conditions combined with all kinds of mathematic optimal technologies.

In this chapter, basic magnet principles, methods of generating a magnetic field, magnetic field applications and numerical methods for the magnet structure design are briefly introduced and reviewed. In addition, the main numerical optimal technology is introduced.

2. Magnet classification

A magnet is a material or object which produces a magnet field. Magnets can be classified as permanent magnets and electromagnets. A permanent magnet is made of magnetic material blocks, has a simple structure and lower costs. However, the magnetic field strength produced by permanent magnets is weak. Electromagnets can operate under steady-state conditions or in a transient (pulse) mode and electromagnets can also be subdivided into resistance and superconducting magnet. A resistance magnet is usually solenoid wound by resistance conductors normally with cooper or aluminum wires and the magnetic field strength is also relatively weaker, but larger than the field generated by permanent magnet. The volume, however, is huge and the magnet system needs a cooling system to transfer the heat generated by the coils' Joule heat. A superconducting magnet is wound by superconducting wires and there is almost no power dissipation due to the zero resistance characteristics of superconductors. The magnetic field strength generated by a superconducting magnet is strong, but limited by the critical parameters of the particular superconducting material. Scientists are trying to improve the performance of superconductors in order to construct superconducting magnets with high critical current density and low operating temperature.

3. Applied superconducting magnet

With the development of superconducting magnets and cryogenic technology, the magnetic field strength of superconducting magnet systems is increasing. A high magnetic field can provide technical support for scientific research, industrial production, medical imaging, electrical power, energy technology etc. Up to now, magnetic fields of about 23 T have been mainly based on low temperature superconductors (LTS), such as NbTi, Nb_3Sn, and/or Al_3Sn. Superconducting magnets with a magnetic field of 35 T are operated in superfluid helium combined with a high temperature superconductor operated at 4.2 K. Magnets with magnetic fields above 40 T are hybrid magnets, consisting of a conventional Bitter magnet and a LTS magnet. Superconducting magnets based on the second generation of YBCO high temperature superconductors may produce a 26.8-35 T magnetic field, while a magnetic field of up to 25 T is possible based on Bi2212 and Bi2223 superconducting magnets. Therefore, research on high magnetic field applications based on superconducting magnet technology has already reached a relatively mature stage.

3.1. Magnet in energy science

With the global growth of economics and an ever increasing population, energy requirements have been growing fast. Up to now, the available sources of energy around the world are nuclear fission, coal, petroleum, natural gas and various forms of renewable energy. Fusion energy has great potential to replace traditional energy in the future because it is clean and economical. The magnetic field is used to balance the plasma pressure and to confine the plasma. The main magnetic confinement devices are the tokomak, the stellarator and the magnetic mirror, as well as the levitated dipole experiment (LDX). Tokamak has become the most popular thermonuclear fusion device in all countries around the world since the Soviet Tokamak T-3 made a significant breakthrough on the limitation of plasma confined time. The magnetic field strength should be strong enough for the fusion energy to be converted to power and superconducting magnet technology is the best solution to achieve high field strength. The superconducting magnet system of Tokamak consists of Toroidal Field (TF) Coils, Poloidal Field (PF) Coils and Correction Coils (CC) (Peide Weng et al., 2006). There are several famous large devices including T-3, T-7 and T-15 in Russia, EAST in China, KSTAR in Korea, JT-60SC in Japan, and JET in UK which have been developed and ITER in France will be installed in the future. Fig. 1 illustrates the main technical parameters for the development of some fusion devices.

Figure 1. The technical parameters for the development of some fusion devices

A magnetohydrodynamics (MHD) generator is an approach to coal-fired power generation with significant efficiency and lower emissions than the conventional coal-fired power plant. The MHD-steam combined cycle power plant could increase the efficiency up to 50-60%, which will result in a fuel saving of about 35%. Its applications could provide great potential in improving coal-fired electrical power production. Since the middle of the 1970s, MHD superconducting magnet development has been ongoing and a series of model saddle magnets have been designed, constructed, and tested (Luguang Yan, 1987).

With the commercialization of high temperature superconductors (HTS), various countries and high-tech companies have made great efforts to strengthen their investment in research on superconductivity, and HTS applications have developed rapidly from 1986. At present, HTS cables, current limiters, transformers, and electric motors have already entered the demonstration phase, while experimental prototypes for HTS magnetic energy storage systems have already appeared. Superconducting Magnetic Energy Storage (SMES) technology is needed to improve power quality by preventing and reducing the impact of short-duration power disturbances. In a SMES system, energy is stored within a superconducting magnet that is capable of releasing megawatts of power within a fraction of a cycle to avoid a sudden loss of line power. SMES has branched out from its original application of load leveling to improving power quality for utility, industrial, commercial and other applications. In recent years superconducting SMES systems equipped with HTS have been developed. A HTS magnet with solid nitrogen protection was developed and used for high power SMES in 2007 by IEECAS (Qiuliang Wang et al., 2008), and 1 MJ/0.5 MVA HTS SMES was developed and put into operation in a live power grid of 10 kV in late 2006 at a substation in the suburb of Beijing, China (Liye Xiao et al., 2008). The LTS magnet fabricated with compact structure for 2 MJ SMES consists of 4 parallel solenoids to obtain good electromagnetic compatibility for the special applications. The SMES are shown in Fig. 2 (Qiuliang Wang et al., 2010).

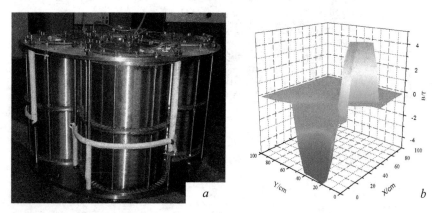

Figure 2. (*a*) The magnets (*a*) and (b) The magnetic field distribution 2 MJ SMES

3.2. Ultra-high superconducting magnet in condensed physics

In order to develop a 25-30 T complete high magnetic field superconducting magnet with an HTS magnet system, NHMFL and Oxford Superconductivity Technology (OST) established a collaboration to develop a 5 T high temperature superconducting insert combined with a water-cooled magnet system. They achieved a central field of 25 T in August 2003 (H. W. Weijer et al., 2003). By using an YBCO HTS magnet as an insert coil in 2008, the total field was increased to 32.1 T, and a 35.4 T layer-wound YBCO magnet has subsequently been fabricated and tested. The German Institut für Technische Physik (ITEP) at the Karlsruhe Institute for Technology (KIT) (M. Beckenbach et al, 2005) used Bi2223 to successfully develop a 5 T insert coil, which operates under a 20 T background magnetic field. The development of this technology provided the technological basis for the development of a high field NMR system. The low temperature required to operate a 20 K HTS magnet can be obtained through a Gifford-McMahon (GM) refrigerator. Because the specific heat at 20 K increases about by two orders of magnitude compared with that at 4.2 K, HTS magnets have higher stability compared with LTS. The HTS magnets with fields of 3.2–5 T were developed and operated as insert coils in a 8-10 T/100 mm split-pair system in China (Yinming Dai et al., 2010), the configuration is shown in Fig. 3. The largest HTS magnet project in that laboratory is focused on developing a 1 GHz insert coil (W. Denis Markiewicz et al., 2006). Although the field threshold of Bi2223 and Bi2212 HTS tapes is over 30 T, operation with HTS tapes is limited due to the Lorentz forces. In order to obtain stable HTS magnets, the persistent current mode is used for HTS inserts, with the aim of obtaining field stability smaller than 10^{-8}/h and field uniformity below 10^{-9} in the region of Φ10 mm × 20 mm. The solenoid-type configuration has more advantages than the double pancake structure.

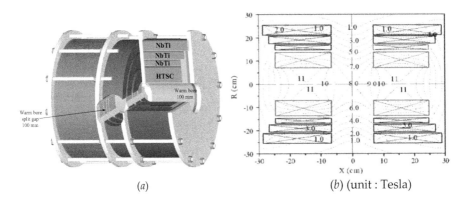

(a)   (b) (unit : Tesla)

Figure 3. Configuration of 8-10 T/100 mm split-pair (a) and (b) The field distribution

The 40 T hybrid magnet system will be designed and constructed at the High Magnet Field Laboratory, Chinese Academy of Sciences (HMFLCAS), and the construction of the hybrid magnet is planned to be completed in 2013. The hybrid magnet consists of a resistive insert providing 29 T and a superconducting coil providing 11 T on the axis over a 32 mm bore (W. G. Chen et al., 2010). The outsert with 580 mm room temperature bore consists of two sub-coils, the inner one (coil C) is a layer wound of Nb₃Sn conductor and the outer one (coil D) is a layer wound of NbTi conductor. Both conductors adopt a cable-in-conduit conductor and will be cooled by 4.5 K force-flowed supercritical helium. For the future upgrade, two Nb₃Sn sub-coils (coil A and coil B) will be inserted into the 11 T superconducting outsert coils and the maximum field in the superconducting magnet will be more than 14 T. Moreover, the resistive insert will be upgraded to 31 T and the total system central field will be above 45 T. Fig.4 shows the overall configuration and a cross-section of the outsert of the 40 T hybrid magnet system.

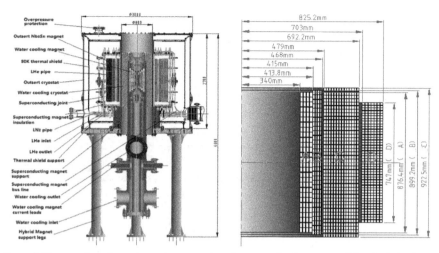

Figure 4. The overall configuration and cross-section of superconducting outsert of the 40 T hybrid magnet system at HMFLCAS

3.3. Magnet in NMR, MRI and MSS

Since the first Nuclear Magnetic Resonance (NMR) spectrometer magnet system was invented in 1950, NMR has been widely used in leading laboratories all over the world as an effective tool for materials research and it has become the most important analysis tool for modern biomedicine, chemistry and materials science. The use of a superconducting coil for the NMR system (instead of a resistive one) has the advantages of low energy consumption, compact coil structure, stable current and magnetic field, good field uniformity, and high magnetic field. Appropriate superconductors for high field application are now Nb₃Sn or the ternary compound (NbTa)₃Sn. HTS materials, such as YBCO and Bi2212, will be the main superconductors in the future. At present, the standard NMR magnet has an aperture of 52 mm and the magnetic field range is from 4.7 T to 23.5 T. The corresponding frequency is between 200 and 1000 MHz, and the stored energy ranges from 18 kJ to 26 MJ (Bernd Seeber, 1998). High field NMR systems need field stability better than 10^{-8}/h and a magnetic field uniformity of 2×10^{-10} in a 0.2 cm³ spherical volume. In 2010, the Bruker Corporation developed a 1000 MHz LTS NMR spectrometer, demonstrating that the LTS conductors NbTi and Nb₃Sn have reached their limit. A 400MHz NMR superconducting magnet system was designed, fabricated and tested at IEECAS (Qiuliang Wang et al. 2011). To meet the requirements of 400 MHz high magnetic field nuclear magnetic resonance, the superconducting magnets are fabricated with 17 coils with various diameters of superconducting wire to improve the performance and reduce the weight of the magnet. In order to reduce the liquid helium evaporation, a two-stage 4 K pulse tube refrigerator is employed. The superconducting magnet with available bore of Φ54 mm is shown in Fig.5.

Figure 5. Configuration of 400 MHz superconducting magnet with cryostat

Since 1980, magnetic resonance imaging system (MRI) magnet technology has made continuous progress in medical diagnosis. In the past 30 years, MRI has developed into one of the most important medical diagnosis tools. Due to the clear soft tissue imaging, MRI technology maintains its leading status in medical applications. The key issue in designing and constructing a MRI superconducting magnet is obtaining a highly uniform and stable magnetic field over an imaging volume. The trend in MRI development, therefore, is toward short length of coils, high magnetic field and a fully open, rather than tunnel-like, structure. The shortest coil length up to now is 1.25 m to reduce the patient's incarceration sickness and achieve lower helium consumption.

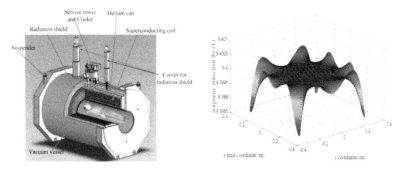

Figure 6. Configuration of cryostat and the field distribution over the DSV region

At present, the designs of open-style MRI systems use permanent (field range from 0.35 T to 0.5 T) or superconducting magnets. Magnets with fields below 0.7 T can use the combination of a superconducting coil and an iron yoke, which produces a highly uniform field. Standard clinical 1.5 T and 3 T MRI scanners have developed rapidly and now installed in many hospitals. The higher filed devices, may be 7 T MRI, will become the next generation clinical scanner and are supported by three big commercial companies (GE, Philips, and

Siemens). The first 7 T whole body scanner with passive shielding was installed in 1999. The first actively shielding 7 T device was designed by Varian and Bruker will soon launch a similar one. The first 9.4 T, which is equivalent to 400MHz functional MRI, was manufactured in 2003 by Magnex Scientific Ltd, a company which was incorporated into Varian. An 11.75 T/900 mm superconducting magnet system is in the process of being fabricated in France; it will be used in neuroscience research in the Commissariat à l'Ènergie Atomique (CEA) in France. Since 2011, a 9.4 T superconducting magnet for metabolic imaging has been undergoing development in the Institute of Electrical Engineering, Chinese Academy of Sciences (IEECAS) (Qiuliang Wang et al. 2012). The magnet has a warm bore that is 800 mm in diameter and cryogenics with zero boil-off of liquid helium will be used for cooling the superconductors. The overall configuration and the field distribution over the DSV region are shown in Fig.6, respectively.

A magnetic surgery system (MSS) (Qiuliang Wang et al. 2007) is a unique medical device designed to deliver drugs and other therapies directly into deep brain tissues. This approach uses superconducting coils to manipulate a small permanent magnet pellet attached to a catheter through the brain tissues. The movement of the small pellet is controlled by a remote computer and displayed on a fluoroscopic imaging system. The magnets of the previous generations were composed of three pairs of orthogonal superconducting solenoid coils. The control strategies are complex because of the magnetic field distribution of solenoids. A novel type of spherical coils can generate linear gradient field over a large spherical volume. This type of modified spherical coils with a constant current distribution model is easy to fabricate in engineering. A prototype of this spherical magnet has already been constructed with copper conductors. According to the key research problems of MSS and the disadvantages of the current MSS, we present a novel type of superconducting magnets structure. The first domestic model MSS has also been constructed and a series of experiments have been performed to simulate the real operation situations on this basis.

3.4. Magnet in scientific instrument and industry

Initially, superconducting magnets were used as scientific instruments in laboratory. With the improvement of magnetic field strength and performance, superconducting magnet technology has been applied in many fields such as accelerators, industry and so on.

Accelerators are the most important tools for high energy physics research. The investment costs of the accelerator rings are determined by the ring size and the operating costs by the power consumption of the magnets. Superconducting magnets with high current density and lower costs were widely applied to accelerator fields. There are several large, famous accelerators equipped with superconducting magnets such as Tevatron at Fermilab, Hadron Elektron Ringanlage (HERA) at the Deutsches Elektronen-Synchroton (DESY) and the Large Hadron Collider (LHC) at the European Organization for Nuclear Research (CERN). The accelerator superconducting magnet system includes dipole magnets for particle deflection, quadrupoles for particle focusing and sextupole and octupole magnets for correction purposes. It is difficult to design and construct this magnet system because the field distribution is more complex and all magnets need a high effective current density to get the high field strength. In addition, the required high field quality, which means uniformity in the case of a dipole and exact gradient in the case of a quadrupole, and the required repeatability for the series of magnets operated in a high radiation environment are challenges in the design and construction of these magnets.

The Fermilab Tevatron Proton-Antiproton Collider (D. McGinnis. 2005) is the highest energy hadron collider in the world. The superconducting Tevatron dipole magnet has a magnetic length of 6.116 m and a radial mechanical aperture of 0.0381 m. The coil package is assembled with an upper coil and a lower coil each of which has an inner layer of 35 turns and an outer layer of 21 turns. The Rutherford-style cable is composed of 23 strands, 12 coated with ebanol and 11 with Stabrite. Each of these strands has 2050 NbTi filaments with the diameter of about 9 μm. The filament separation to diameter ratio is 0.35 and the ratio of copper to non-cooper is 1.8. The coil package is enclosed in a cylindrical cryostat inserted into a warm iron yoke.

The HERA (R. Meinke. 1991) installed at DESY consists of 650 superconducting main magnets (dipoles and quadrupoles) and approximately the same number of superconducting correcting elements (dipoles, quadrupoles and sextupoles). The system consists of two independent accelerators designed to store 30 GeV electrons and 820 GeV protons, respectively. These magnets formed a continuous cold string through the 6.3 km long HERA tunnel interrupted only by warm sections around the interaction regions. The superconducting dipoles with the central field of 4.68 T and the magnetic length of 8.824 m, and the superconducting quadrupoles with the central gradient field of 91.2 T/m and the magnetic length of 1.861 m are of the cold bore and cold yoke type.

The LHC (L. Rossi. 2003) is a gigantic scientific instrument near Geneva, where it spans the border between Switzerland and France about 100 m underground. It is a particle accelerator used by physicists to study the smallest known particles – the fundamental building blocks of all things. Most of its 27 km underground tunnel was filled with superconducting magnets, mainly 15 m long dipoles and 3 m long quadrupoles. The LHC magnets are operated at the field strength of 8.36 T at an operating temperature of 1.9 K, which is approaching the 11.45 T mark that is considered to be the upper limit for a niobium-titanium superconductor. In the LHC accelerator, the stronger the magnetic field is, the tighter the arc of the beam is in its 27 km tunnel. With stronger dipole magnets, an accelerator can push particles to much higher relativistic energies around the same-sized circular beam path.

The Superconducting Solenoid Magnet (SSM) is designed to provide a uniform 1.0 T axial field in a warm volume of 2.75 m diameter. It is the first superconducting magnet of this type built in China. The 0.7 mm diameter NbTi/Cu strands are formed into a Rutherford cable sized 1.26 mm × 4.2 mm. The Rutherford cable is embedded in the center of a stabilizer made of high purity (99.998 %, RRR 500) aluminum with outer dimensions of 3.7 mm × 20.0 mm. One layer of 0.075 mm thick Upilex-Glassfibre (glass fiber reinforced polyimide) film is used for turn-to-turn insulation of the coil winding. The superconducting magnet is indirectly cooled by a forced flow of two phase helium at an operating temperature of 4.5 K through cooling tubes wound on the outside surface of the support cylinder (B. Wang et al. 2005). The Superconducting IR Magnets (SIM) for the BEPC upgrade (BEPC-II) are installed completely inside the BES-III detector and operated in the detector solenoid field of 1.0 T. In BEPCII, a pair of superconducting quadrupole magnets (SCQ) (L. Wang et al. 2008) with high focusing strength is used, which will squeeze the β function at the interaction point and provide a strong and adjustable magnet field. Both of the two SCQs are inserted into the BESIII detector symmetrically with respect to the interaction to produce an axial steady magnetic field of 1.0~1.2 T over the tracking volume and to meet the requirements of particle momentum resolution to particle detectors. Fig.7 shows the components and fabrication at the site of the SCQ.

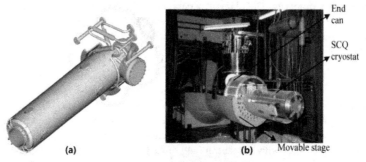

Figure 7. (a) SCQ system components and (b) SCQ fabrication on site

The MICE coupling magnet (D. Li et al. 2005) consists of a single 285 mm long superconducting solenoid coil developed by the Harbin Institute of Technology. The superconducting coil is wound on a 6061 aluminum mandrel that is fitted into a cryostat vacuum vessel. The inner radius of the coil is 750 mm and its thickness is 110 mm at room temperature. The coil assembly is comprised of the coil with electrical insulation and epoxy, and the coil case is made of 6061-T6-Al, including the mandrel, end plates, banding, and cover plate. The length of the coil case is 329 mm. The coupling solenoid will be powered by a single 300 A/0–20 V power supply connected to the magnet through a single pair of leads that are designed to carry a maximum current of 250 A. It is cooled by liquid helium flow through cooling tubes embedded in the coil cover plate by two 1.5 W cryocoolers.

Figure 8. (a) Cryostat and (b) Prototype of dipole magnet for GSI CR ring.

The super-ferric dipole prototype of the Super Fragment Separator (Super-FRS) has a width of 2200 mm, a central length of 2020 mm and a height of 725 mm, respectively. The Collector Ring (CR) magnet has been built by the Facility for Antiproton and Ion Research (FAIR) China Group (FCG), including IMPCAS, IPPCAS, and IEECAS, in cooperation with the GSI Helmholtz Center for Heavy Ion Research (GSI) (Hanno Lei brock et al. 2010). IEECAS is contributing to the design of the coils, IPPCAS is responsible for the fabrication of the superconducting coils and cryostat and IMPCAS is responsible for the testing of the whole system, the magnetic field optimization and the development of the 50 ton laminated iron yoke. The dipole magnet has a homogeneous region 380 mm in width and 140 mm in height, while the homogeneity reaches $\pm 3 \times 10^{-4}$. The passive air slot and chamfered removable end poles guarantee that the magnetic field distribution is homogeneous at both low and high field levels. The Super-FRS superconducting dipole is a super-ferric superconducting magnet with a warm iron yoke which is laminated due to magnetic field ramping and the H type yoke is made of laminated electrical steel 0.5 mm in thickness, which was stamped and glued to blocks, and machined to the 15 degree angle sector shape. The superconducting coils were wound from multi-filamentary NbTi wires with a higher than usual the ratio of copper to non-copper and are operated at liquid helium temperature. The coils are positioned in the helium cryostat with a multi-layer insulation structure. The total weight of the magnet is more than 52 tons. The magnetic field measurements indicate that the field homogeneity is about $\pm 2 \times 10^{-4}$ at different magnetic field levels (0.16 T - 1.6 T), which is better than the design requirements. The cryostat at the test facility in IPPCAS and prototype dipole at the test facility of IMPCAS are shown in Fig. 8.

For the requirements of microwave devices, a conduction-cooled magnet has been fabricated for the microwave experiments used in Gyrotron. A magnet system with a center field of 1.3–9 T and warm bore of Φ 100 mm has been designed and fabricated (Qiuliang Wang et al. 2007). The electromagnetic structure of the magnet is designed on the basis of the hybrid genetic optimal method. The length of homogeneous region of the superconducting magnet is adjustable from 200 mm to 250 mm. Also the superconducting magnet can generate multi-homogeneous regions with the length of 200, 250 and 320 mm. The homogeneity of the magnetic field is about $\pm 0.5\%$ with a constant homogenous length and $\pm 1.0\%$ for adjusting homogenous length. All of the homogeneous regions start at the same point and the field decays to 1/15-1/20 from the front point of a homogeneous region to 200 mm. The superconducting magnet is cooled by one GM refrigerator with cooling power of 1.5 W at 4 K. The configuration of the superconducting magnet with superconducting coils and copper coils is illustrated in Fig.9 (a). The homogeneous region length of 320 mm with maximum center field of 1.3T is generated by main coils 1, 2 and compensating coils 3 and 5. In order to extend the magnetic field decay from 200 mm to 300 mm, we need to use the normal copper coils. Therefore, the homogeneous region length of 320mm with a field of 1.3 T is generated by the main and compensating coils of 1, 2, 3, 5, 6 and 7 for superconducting coils and 8, 9 and 10 for copper coils. The homogeneous region length of 250 mm with the field of 4 T, and the magnetic field decay can be adjustable through the main and compensating coils of 1, 2, 3, 5, 6 and 7. The magnetic field decay can be controlled through the adding cathode compensation coils of 6 and 7, and the coils are connected to an assisting power supply to adjust the operating current. The total superconducting coil set-up should have five high temperature superconducting current leads. The copper adjustment coils 8, 9, and 10 are used to change the operating current to correct the magnetic field distribution in the homogeneous region. The main field distributions are illustrated in Fig.9 (b). For the requirements of IEECAS customers, superconducting magnets with all kinds of magnetic field distribution are fabricated.

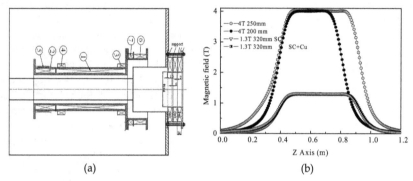

(a)   (b)

Figure 9. Superconducting magnet with 10 coils arranged with the same axis: the superconducting coils are installed in the cryostat; the copper coils are located outside of the cryostat and fixed on its flange, where they are cooled by air convection. The superconducting coils are cooled by a GM cryocooler (a), Magnetic field distribution for various lengths of homogeneous region (b).

4. Structure and function of magnets

The desired magnetic field produced by superconducting coils and the shape of field is predetermined by the users and its special application. The magnetic field distribution depends on the size and shape of coils and final system structure. The common shapes of superconducting coils are solenoid, saddle coils, race-track coils, toroid coils, baseball coils and yin-yang coils for different applications.

4.1. Configuration of solenoid magnet

The most efficient and economic coil is the solenoid structure, and normal solenoids are symmetric consisting of a single solenoid or several coaxial solenoids based on the field distribution and homogeneity demands. The solenoid coil is wound layer by layer with round or rectangular cross-section wires on a cylindrical bobbin. The basic parameters for a solenoid are inner radius r_{inner}, router radius r_{outer}, the length L and the current density J. The current density $J = NI_{op}/[L(r_{outer}-r_{inner})]$, where the number of windings and operating current are N and I_{op}, respectively. The conductor current density is higher due to the electrical insulation and the eventual mechanical reinforcement. By these parameters, the magnetic field can be calculated by the popular equation (Martin N. Wilson. 1983). By this method, in theory, we can design all symmetric field distribution magnet system.

4.2. Racetrack and saddle-shaped magnet

The racetrack-shaped coil has two linear segments and two semicircular arc segments. The saddle coil has two linear segments and six small circular segments. The coil structure of racetrack-shaped and saddle coils are shown in Fig. 10. The racetrack-shaped magnet may be used in electrical machinery, magnetic levitation trains, dipole or multipole magnets for an accelerator, wiggler and undulator magnets, large-scale MHD magnets, space detector magnets and in space astronaut radiation shield and accelerator detector magnets, such as the ATLAS magnet at CERN. Sometimes, accelerator magnets, electrical machinery magnets and MHD magnets employ saddle shaped coils. Transverse magnetic field distribution can be produced by combining with the saddle coils and change the current direction. Saddle coils are also used to correcting the magnetic field distribution for magnetic resonance magnets and magnetohydrodynamics.

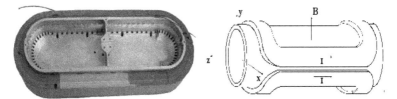

Figure 10. Configurations of a racetrack magnet and a saddle magnet.

4.3. Structure of other complicated magnet

Baseball coils and yin-yang coils is used to confine the plasma as magnetic mirror. The baseball coils with U-shaped structure produce a magnetic field magnitude increasing in every direction outwards from the plasma and the structure is more economic than the same mirror field produced by a pair of solenoid. A yin-yang magnet consists of two orthogonal baseball coils which generally produce a deeper magnetic well than a single baseball coils and also use fewer conductors. The magnet structure of a magnetic mirror is even more complex, as for the stellarator shown in Fig.11 (a). The magnet current distribution forms a yin-yang structure. Force-free magnets are ones in which the current density J is parallel to the field H everywhere, i.e. $J = \alpha H$, where α is a scale function called the force-free function or factor. The Lorentz force f is therefore equal to zero since $f = \mu J \times H = 0$. However, from the virtual work theorem of mechanics, it can be verified that it is impossible to be force-free everywhere in a finite electromagnetic system without magnetic coupling with other systems (Yanfang Bi & Luguang Yan. 1983). Furthermore, it is also unnecessary to construct a fully force-free magnet, as shown in Fig.11 (b), since the solid coil itself could withstand certain forces. So we need practically to develop some quasi-force-free magnets in which J and H are approximately parallel, so that although the Lorentz forces are not zero, they are reduced significantly. With the development of accelerator magnet technology in recent years, the so-called snake-shaped dipole magnet, shown in Fig.11(c), has been proposed, which can deliver good magnetic field uniformity. This kind of magnet can be used in accelerators for particle focusing.

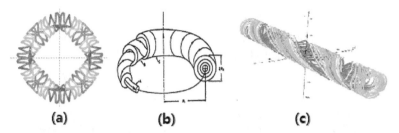

(a)   **(b)**   **(c)**

Figure 11. Configuration of (*a*) stellarator, (*b*) force-free, and (*c*) snake-shaped magnets.

5. Numerical methods for magnet design

Due to the complex structure of electromagnetic devices and the compact design requirements, the design of modern magnets no longer relies on simple analytical calculations. Usually, the designers employ complex high-level numerical analysis technology to decide the electromagnetic structure parameters. With the geometry, the material distribution and the driven sources given, the numerical analysis of the electromagnetic field distribution with respect to space and time can be conducted by solving the Maxwell's equations numerically under predefined initial and boundary conditions. During the design of magnet devices, the designer should propose a configuration satisfying the functional needs as far as possible. The inverse problem is: given the magnetic field distribution in space and time, one must find the geometric parameters and the material distribution, as well as the field source (K. Huang & X. Zhao. 2005). Magnet design is such an electromagnetic field inverse problem. Its task is to find the field source (current distribution or permanent magnet material distribution) on the basis of a given magnetic field spatial distribution. The inverse problem has two different aspects: the design optimization itself and the parameter identification. (Ye Bai et al. 2006; W. Zhang, H. He & A. P. Len. 2005).

The deterministic method is doing the search gradually during the iterative process according to the search direction determined by each step of the iteration, so that the objective function value of the current step iterative solution is certain to be smaller than the preceding values. Different deterministic methods have different search directions, such as the "steepest decent", the "conjugating gradient method", the "Quasi-Newton law" (Chunzhong Wang, Q. Wang & Q. Zhang. 2010), the "Levenberg-Marquard algorithm", etc. The deterministic method depends on the neighborhood characteristics of the current search position to determine the next step search position (with partial linearization in a non-linear problem). Therefore it is local optimization and the efficiency of seeking for the local optimal solution is very high, but it does not have global optimization capability in a multi-extreme value problem. Another shortcoming of the deterministic method is that it is necessary to know the first- or second-order partial derivative of the objective function and it usually requires the objective function to not be too complex and to have an analytic expression, which increases the computing time and cost. On the other hand, the ill-posed inverse problem is often inherited with an optimization problem. Therefore, the regularization processing should be added to each iteration step of the deterministic method, as otherwise big errors will occur, and the iteration may not work.

In order to avoid the limits of the deterministic method, the stochastic method (Monte Carlo) is suggested (N. Metropolis & S. M. Ulam. 1949). The Monte Carlo method works in such a way that each iteration step is determined by a random number. The traditional Monte Carlo method carries on a completely stochastic blind search, assuming that all possible solutions have equal probability. In contrast, the modern Monte Carlo methods, such as the well-known simulated annealing method (S. C. Kirkpatrick, D. Gelatt & M. P. Vecchi. 1983), the genetic algorithm (Qiuliang Wang et al. 2009), the evolutionary algorithm (ant colony algorithm and particle swarm optimization), the taboo search method, and the neural network and other stochastic algorithms, carry on the random search in a more instructive way, giving the different possible solutions with different probabilities. The merits of the Monte Carlo method are: it is universally serviceable and no target problems need to be differentiated as to whether they are linear or non-linear, ill-posed or well-posed. A problem can be processed by the Monte Carlo method, even if its operator is very complex and cannot be expressed with an analytical formula. Besides, the method has a strong optimization capability in all situations. Its shortcoming is that the calculation time is usually very large, growing inordinately with the order of the problem, while the convergence rate is very slow.

In order to combine the respective merits of the above algorithms, many researchers have been striving to work for the unification of these methods. In order to reduce the computing time, a new kind of optimizing strategy has emerged in recent years – the unification of the response surface model and the stochastic optimized algorithm (J. H. Holland. 1992; C.W. Trowbridge. 1991). This method firstly separates the space of the target variable for a series of sampling points and then applies the numerical calculus method to compute the value of the objective function on these sampling points; with these values, it uses a response model to reconstruct the objective function and then the optimization computation is carried out using the optimizing algorithm on the restructured objective function. Because it is only necessary to calculate the value of the electromagnetic field objective function on the sampling points, the algorithm efficiency is enhanced greatly. Sometimes in the optimization design of an electromagnetic installation, unifying the Moving Least Squares method with the simulated annealing method has very good results (Chao Wang et al. 2006). The convergence rate of these algorithms, however, still cannot satisfy the requirements for computing complex large-scale systems, for example, three-dimensional calculations, transient processes or coupled systems, at present. In magnet design, a combination of the deterministic and the stochastic algorithms has been adopted.

6. Design example of high homogeneous magnet

A high homogeneous magnet system is the most important and expensive component in an MRI or NMR system. A superconducting magnet with the distribution of coils in single layer, two layers and even more layers is the best solution to achieve the high magnetic field strength and homogeneity requirements. The challenge for designing a MRI magnet system is to search the positions and sizes of coils to meet the field strength and homogeneity over the interesting volume and stray field limitation.

6.1. Mathematical model for a hybrid optimization algorithm

The parameters of the system, including length, inner and outer radius of feasible region of coils, radius of DSV region and the axial and radial radius of 5 Gauss stray field line, are predetermined by designer based on the actual applications. Fig. 12 illustrates an example for the design of a symmetric solenoid magnet system. The required parameters of the feasible rectangular region for arrange coils are inner and outer radius (r_{min}, r_{max}) and the length (L), the interesting imaging volume is commonly sphere and the stray field is limited to smaller than 5 Gauss outside the scope of an ellipse. A hybrid optimization algorithm by combination of Linear Programming (LP) and Nonlinear Programming (NLP) was developed by IEECAS. This approach is very flexible and efficient for designing any symmetric and asymmetric solenoid magnet system with any filed distribution over any shape volume.

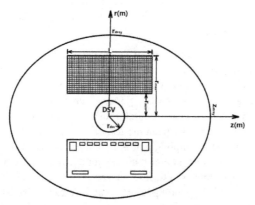

Figure 12. The region of feasible coils, interesting volume and stray field

Firstly, the feasible rectangular region can be meshed as 2-D continuous elements with N_r elements for radial direction and N_z elements for axial direction, and each element served as an ideal current loop. The surface of the sphere and the ellipse for homogenous filed and stray field limitation were evenly divided into N_d and N_s parts along the elevator from 0 to π, respectively. The field distribution at all target points including N_d and N_s points produced by all ideal current loops with unit current amplitude calculated, and the unit current field contribution matrices A_d and A_s were formed. A LP algorithm was built up with the objective functions totaling the volume of superconducting wires, the field distributions at all target points and the maximum current aptitude for all current elements were constrained. The LP mathematical model is formulated as following:

$$\text{Minimum: } \sum_{i=1}^{Nz+Nr} r_i |I_i| \tag{1}$$

$$\text{Subject to: } \begin{cases} \left| A_d * I - B_0 \right| / B_0 \le \varepsilon \\ \left| A_s * I \right| \le 5 Gauss \\ I \le I_{max} \end{cases} \tag{2}$$

The current map with sparse nonzero clusters were calculated by first LP stage, the positions of nonzero clusters can be discretized into several solenoids with the size of inner radius (r_{inner}), outer radius (r_{outer}), and the z position of two ends (z_{left}, z_{right}). Secondly, a NLP algorithm was built up, and the objective function and constraints of the algorithm are similar to the LP stage, and added current margin constraint into the algorithm which based on the maximum magnetic field within the superconducting coils and the relationship of critical current and magnetic field. The NLP mathematical model is formulated as following:

$$\text{Minimum: } \sum_{i=1}^{Ncoils} V_i \tag{3}$$

$$\text{Subject to: } \begin{cases} \left[\max(B_{zdsv}) - \min(B_{zdsv}) \right] / mean(B_{zdsv}) \le H \\ \sqrt{B_{zstray}^2 + B_{rstray}^2} \le 5 Gauss \\ I / Ic(B_{max}) \le \eta \end{cases} \tag{4}$$

here, the $Ncoils$ is the number of discretized solenoids, V_i is the volume of the ith solenoid, B_{zdsv}, B_{zstray} and B_{rstray} are the magnetic field on DSV region and stray field ellipse region, respectively. η is the current margin, B_0 is the target field over the DSV and H is the homogeneity level for design.

Many cases were studied by this hybrid algorithm, and the results show the method is flexible and efficient for the first LP stage which took about 5 minutes and the second NLP stage which took about 30 minutes, respectively. The resultant coil distributions were simple and easy to fabricate.

6.2. Design cases

6.2.1. 1.5 T actively shielded symmetric solenoid MRI

The actively shielded symmetric MRI system with the length of 1.15 m, the central magnetic field of 1.5 T and the field quality of 10 ppm over 500 mm DSV and the radial inner and outer radius of 0.40m and 0.80 m, the stray field of 5 Gauss outside the scope of an ellipse with axial radius of 5 m and radial radius of 4 m, and the current margin of 0.8. The N_r, N_z, N_d, and N_s were set as 40, 40, 51 and 51, respectively. The current map by LP and the final actual coils sizes and positions are shown in Fig. 13. The coils distribution with two layers, the inner layer with four pairs of positive and one pair of negative current direction coils for producing the required magnetic strength and the homogeneity. The outer layer with a pair of negative current direction coils for reduces the stray field strength. The homogeneities and stray field distributions are shown in Fig. 14. The coils parameters are shown in Table I. The operating current density and actual current margin are 148 MA/m^2 and 0.7546, the maximum magnetic field and hoop stress are 5.43 T and 145.16 MPa, respectively.

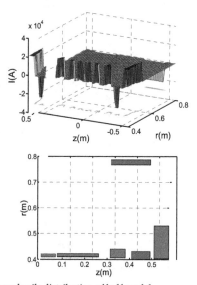

Figure 13. The current map and coils distribution of half model

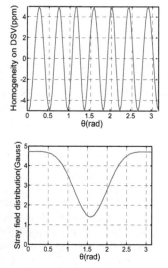

Figure 14. The homogeneity over 500 mm DSV and stray field distribution

6.2.2. 1.0 T open biplanar MRI

The unshielded open biplanar MRI system has a central field strength of 1.0 T, inner and outer radius of 0.0 and 0.90m, lower and upper z positions for two ends of 0.40m and 0.45 m, and field quality of 15ppm over 450 mm DSV.

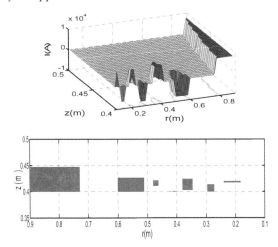

Figure 15. A quarter model current map and coils distribution

The N_r, N_z, N_d, and N_s were set as the same as the design of 1.5 T actively shielded MRI system. A quarter model current map for the LP stage and the coils distribution are shown in Fig. 15. The coils distributions with single layer have six coils, the largest coils with maximum field of 5.26 T are the outermost coils with positive current direction. The coils parameters are shown in Table II. The operating current density and actual current margin are 150 MA/m^2 and 0.74, the hoop stress is 128.2 MPa, respectively.

7. Conclusion

In this chapter we described the basic physical concepts of magnet design and the classification of magnet, then illustrated the magnet applications in the field of energy science, condensed physics, medical devices, scientific instruments and industry. Electromagnetic design is very important for the design of a magnet system with optimal coils distribution. Numerical methods are the best solution to designing complex field distribution magnet systems. A hybrid optimization algorithm combined with linear programming and nonlinear programming was presented and studied.

Author details

Qiuliang Wang, Zhipeng Ni and Chunyan Cui
Institute of Electrical Engineering, Chinese Academy of Sciences, China

8. References

B. Seeber. (1998). *Handbook of Applied Superconductivity*, Taylor & Francis (edition), ISBN 0750303778, UK

B. Wang et al. Design, Development and Fabrication for BESIII Superconducting Muon Detector Solenoid, *IEEE Transactions on Applied Superconductivity*, Vol.15, No. 2, (June 2005), pp. 1263-1266, ISSN 1051-8223

C. W. Trowbridge. (1991). An Introduction to computer aided electromagnetic analysis, Vector Field Ltd., ISBN 0951626205 9780951626207, UK

Chao Wang et al. Electromagnetic optimization design of a HTS magnet using the improved hybrid genetic algorithm, *Cryogenics*, Vol. 46, (2006), pp. 349-353, ISSN 0011-2275

Chunzhong Wang, Q. Wang & Q. Zhang. Multiple Layer Superconducting Magnet Design for Magnetic Resonance Imaging, *IEEE Transactions on Applied Superconductivity*, Vol.20, No. 3, (June 2010), pp. 706-709, ISSN 1051-8223

D. Li et al. (2005). Progress on the RF Coupling Module for the MICE Channel, *Proceedings of 2005 Particle Accelerator Conference*, ISSN 1051-8223, Knoxville, Tennessee, May 2005

D. McGinnis. (2005). Fermilab TEVATRON operational status, *Proceedings of 2005 Particle Accelerator Conference*, ISSN 1051-8223, Knoxville, Tennessee, May 2005

H. W. Weijer et al. Development of 5T HTS insert magnet as part of 25T class magnets, *IEEE Transactions on Applied Superconductivity*, Vol.13, No.2, (June 2003), pp. 1396-1399, ISSN 1051-8223

Hanno Lei brock et al. Prototype of the Superferric Dipoles for the Super-FRS of the FAIR-Project. *IEEE Transactions on Applied Superconductivity*, Vol.20, No. 3, (June 2010), pp. 188-191, ISSN 1051-8223

J. H. Holland. (1992). *Adaptation on Natural and Artificial Systems*. Univ. of Mich. Press, Ann Arbor, Mich., ISBN 0262581116, USA

K. Huang & X. Zhao. *The inverse problem and application of electromagnetic field*, Science Press , ISBN 9787030159120, Beijing China

L. Rossi. (2003). The LHC superconducting magnets, *Proceedings of the 2003 Particle Accelerator Conference*, ISSN 1088-9299, Portland Oregon USA, May 2003

L. Wang et al. Superconducting Magnets and Cooling System in BEPCII, *IEEE Transactions on Applied Superconductivity*, Vol.18, No. 2, (June 2008), pp. 146-149, ISSN 1051-8223

Liye Xiao et al. Fabrication and Tests of a 1MJ HTS Magnet for SMES, *IEEE Transactions on Applied Superconductivity*, Vol.18, No. 2, (June 2008), pp. 770-773, ISSN 1051-8223

Luguang Yan. (1987). Applied superconductivity work at the Institute of Electrical Engineering, Academia Sinica, *Cryogenics*, Vol. 27, No.9, (Sep. 1987), pp. 484-494, ISSN 0011-2275

M. Beckenbach et al. Manufacture and test of a 5T Bi-2223 Insert Coil, *IEEE Transactions on Applied Superconductivity*, Vol.15, No.2, (June 2005), pp. 1484-1487, ISSN 1051-8223

Martin N. Wilson. (1983). *Superconducting Magnets*, Oxford University Press, ISBN 0198548052, New York

N. Metropolis & S. M. Ulam. The Monte Carlo methods. *J. Am. Stat. Assoc.*, Vol. 44, (Sep. 1949), pp. 335-341, ISSN 01621459

Peide Weng et. al. (2006). Recent Development of Magnet Technology in China: Large Devices for Fusion and Other Applications. *IEEE Transactions on Applied Superconductivity*, Vol. 16, No.2, (June 2006), pp. 731 – 738, ISSN 1051-8223

Qiuliang Wang et al. Design of Test of Conduction-Cooled High Homogenous Magnetic Field Superconducting Magnet for Gyrotron, *IEEE Transactions on Applied Superconductivity*, Vol.17, No. 2, (June 2007), pp. 2319-2322, ISSN 1051-8223

Qiuliang Wang et al. Development of high magnetic field superconducting magnet technology and applications in China, *Crogenics*, Vol. 47, (2007), pp. 364-379, ISSN 0011-2275

Qiuliang Wang et al. A 30 kJ Bi2223 High Temperature Superconducting Magnets for SMES with solid-Nitrogen Protection. *IEEE Transactions on Applied Superconductivity*, Vol.18, No. 2, (June 2008), pp. 754-757, ISSN 1051-8223

Qiuliang Wang et al. Design of open high magnetic field MRI superconducting magnet with continuous current and genetic algorithm method, *IEEE Transactions on Applied Superconductivity*, Vol.19, No. 3, (2009), pp. 2289-2292, ISSN 1051-8223

Qiuliang Wang et al. High Magnetic Field Superconducting Magnet for 400 MHz Nuclear Magnetic Resonance Spectrometer, *IEEE Transactions on Applied Superconductivity*, Vol. 21, No.3, (June 2011), pp. 2072 – 2075, ISSN 1051-8223

Qiuliang Wang et al. A Superconducting Magnet System for Whole-body Metabolism Imaging, *IEEE Transactions on Applied Superconductivity*, to be published on June 2012, ISSN 1051-8223

Qiuliang Wang et al. Development of large scale superconducting magnet with very small stray magnetic field for 2 MJ SMES, *IEEE Transactions on Applied Superconductivity*, Vol.20, No. 3, (June 2010), pp. 1352–1355, ISSN 1051-8223

R. Meinke. Superconducting magnet system for HERA, *IEEE Transactions on Magnetics*, Vol. 27, No.2, (March 1991), pp. 1728 – 1734, ISSN 0018-9464

S. C. Kirkpatrick, D. Gelatt & M. P. Vecchi. Optimization by simulated annealing, *Science*, Vol. 220, No. 4598, (May 1983), pp. 671-680, ISSN 00368075

W. Denis Markiewicz et al. Perspective on a Superconducting 30T/1.3GHz NMR Spectrometer Magnet, *IEEE Transactions on Applied Superconductivity*, Vol. 16, No.2, (June 2006), pp.1523-1526, ISSN 1051-8223

W. G. Chen et al. Engineering Design of the Superconducting Outsert for 40 T Hybrid Magnet, *IEEE* Transactions *on Applied Superconductivity*, Vol. 20, No.3, (June 2010), pp. 1920 – 1923, ISSN 1051-8223

W. Zhang, H. He & A. P. Len. (2005). Science Computation Introductory Remarks, Tsinghua University Press, ISBN 730210899, Beijing China

Yanfang Bi & Luguang Yan. A superconducting toroidal magnet with quasi-force-free configuration, IEEE Transactions on Magnetics, Vol. 19, No. 3, (May 1983), pp. 324-327, ISSN 0018-9464

Ye Bai et al. Design methods of the spherical quadrupole magnets and sextupole magnets, IEEE Transactions on Magnetics, Vol. 42, No. 4, (April 2006), pp. 1187-1190, ISSN 0018-9464

Yinming Dai et al. An 8T Superconducting Split Magnet System with Large Crossing Warm Bore, *IEEE Transactions on Applied Superconductivity*, Vol. 20, No.3, (June 2010), pp. 608-611, ISSN 1051-8223

Chapter 7

Magnetic Texturing of High-T$_c$ Superconductors

Laureline Porcar, Patricia de Rango, Daniel Bourgault and Robert Tournier

Additional information is available at the end of the chapter

1. Introduction

The magnetic field is an efficient tool for characterizing materials but also for elaborating them with a determined magnetic and crystallographic microstructure. The main effect of a magnetic field is to align crystallites embedded in the liquid along their easy magnetization axis [1,2]. The texturing is successful when the force exerted on the crystal gives rise to a rotation thanks to the presence of a liquid around the particle. The possibility to texture under a magnetic field is nowadays known and is still studied worldwide. However, in this paper, we aim to focus the attention on the role of the overheating above the liquidus temperature where our experiments and calculations show that nuclei are surviving above this temperature [3-7]. This assumption contradicts the classical nucleation model where intrinsic nuclei are not present above the melting temperature T_m and cannot act as growth nuclei while cooling down the temperature below T_m [8-10]. Furthermore, melts can usually be supercooled below their melting or liquidus temperatures (T_m). The degree of supercooling ($\Delta T-$), measured by the difference between the onset temperature of solidification and T_m, is affected by various factors, including the level of overheating and the time. The relations between the overheating ($\Delta T+$), and the supercooling ($\Delta T-$) were studied in Sn and Sn-Pb [11]. The dependence of ($\Delta T+$) on ($\Delta T-$) is a function of the holding time. It is well known that the cooling rate plays an important role in establishing the degree of supercooling since the nucleation of solid structures and thus solidification requires a certain period of time. Reversibly, melting also takes a certain period of time to destroy the order structures. Furthermore, it is worth noting that as long as some residual solid particles (nuclei) exist above the melting temperature, the energy barrier for the nucleation of crystallization during cooling can be reduced and thus the level of supercooling will be nil. As soon as the solid structures in the melt are completely destroyed, a substantial surface energy barrier exists for the nucleation of solid particles upon cooling. Consequently, each supercooling temperature can be associated with a nucleation time. In congruent material, such as Bi, only few crystals are obtained after an overheating of 80°C [1,7]. Then, the magnetic field can act on these remaining nuclei embedded in the liquid above the melting point.

The magnetic field was successfully used to improve texturing in crystals and alloys. The texturing of several alloys under a magnetic field was carefully studied |12]. It depends on the magnetic properties of the crystallizing nuclei and those of the melt. The mechanical force moment allowing the rotation of crystals along the direction of the magnetic field also depends on the degree of homogeneity of the magnetic field. In a homogeneous field, the texture can be induced in a magnetic isotropic crystal due to the moment of forces deriving from the demagnetizing factor and the shape anisotropy of the nucleus. However, at high temperatures and for low susceptibility, this contribution will be neglected. On the contrary, in an anisotropic crystal, the orientation arises from the anisotropic crystal characterized by an difference of magnetic susceptibility $\Delta\chi$ along two mutually perpendicular axes and the moment of forces can be expressed as:

$$K = \frac{\Delta\chi}{2\mu_0} B^2 V sin2\alpha \qquad (1)$$

where α is the angle between B and the axis with maximum $|\chi|$. Under real conditions, there will be a competition between magnetic orientation forces, viscous forces, convective flows in the melt and interactions between crystals or interaction with the crucible. In some cases, the orienting effect may be limited or even negligible. The window in which the orientation can be induced must be carefully found. For an alloy, Mikelson and al. found that the temperature interval in which the orienting actions of the magnetic field are effective lies below the liquidus line and to 20% of the crystallization interval.

In this article, we will focus on the texturing of high-T_c superconductors under a magnetic field where the main conditions exposed below are taken into account. The main difficulties of superconductors texturing reside in a non congruent melting of the compounds. It is the limiting factor of overheating since the phase diagram is generally complex involving a lot of transformations even above the liquidus temperature. A large overheating above the liquidus will usually lead to the formation of unwanted secondary phases during crystallization and prevent the recombination of the superconducting phase. The latter is usually taken as the critical value for the limit of overheating. The texturing under a magnetic field consist of finding the interval of overheating temperatures preventing the transformation of superconducting-phase nuclei in secondary phases and allowing a sufficient amount of liquid. Usually, the allowable overheating is a dozen of degrees above the melting temperature [7]. As will be shown below, this amount of applied overheating does not destroy the presence of nuclei.

2. Magnetic texturing

2.1. Case of $Y_1Ba_2Cu_3O_{7-\delta}$

Rare earth-$Ba_2Cu_3O_{7-\delta}$ crystals are aligned in a magnetic field at room temperature [13] and also when they are imbedded in liquid silver at high temperature [14]. A sufficient paramagnetic anisotropy remains at temperature superior to 1000 °C and the residual anisotropy energy of the crystals is larger than the energy associated with thermal disordering effects [15]. As mentioned previously, the anisotropy energy must also compensate the viscous force in the surrounding liquid where the crystal is free to rotate in a liquid with low resistance. The temperature window inducing a magnetic orientation lies between 1040°C and 1060°C under atmospheric pressure of O$_2$. Below 1040°C, a solid matrix made up of Y$_1$Ba$_2$Cu$_3$O$_{7-\delta}$ (Y123) and Y$_2$Ba$_1$Cu$_1$O$_5$ (Y211) remains throughout the annealing treatment while the melting of Y123 above 1040°C leads to a liquid containing precipitates of Y211. Grains containing inclusions of this secondary phase are partially aligned below 1040 °C with their ab-plane perpendicular to the direction of the annealing field, indicating that a partial melting of the precursor takes place during processing. Both X-ray and magnetization reveal that a weak orientation occurs. For samples prepared at temperature above 1040°C, the texturing is very efficient. The sample is cut and oriented to reveal faces perpendicular to the direction of the magnetic field. X-ray spectrum is taken as a first indication of the orientation. A measure of the orientation P$_{001}$ is given in Figure 1 as a function of annealing temperature [16].

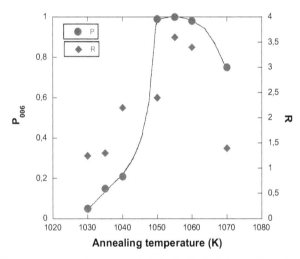

Figure 1. R and P versus annealing temperature in Y$_1$Ba$_2$Cu$_3$O$_{7-\delta}$ bulk textured sample, where R=ΔM$_{H//HA}$/ΔM$_{H\perp HA}$ (ΔM=M$^+$-M$^-$ is the hysteresis in the sample magnetization, M$^+$ and M$^-$ the values of the magnetization measured in an increasing and decreasing field respectively) and P$_{001}$ =1-Γ where Γ=(I$_{hkl}$/I$_{001}$)^o/(I$_{hkl}$/I$_{001}$)^u. I$_{hkl}$ is the intensity of the strongest forbidden non-(00l)line and I$_{001}$ the intensity of a (00l) line in a X-ray diffraction spectrum. The superscripts o and u indicate that measurements were performed on oriented and unoriented samples respectively [16].

In Figure 1, one can note the sharp increase of the orientation for a temperature above 1040°C while the latter gradually decreases above 1050°C. The orientation is deduced from the comparison of R, the hysteresis in sample magnetization ratio between parallel and perpendicular field. R=ΔM$_{H//HA}$/ΔM$_{H\perp HA}$ and ΔM=M$^+$-M$^-$ is the hysteresis in the sample magnetization where M$^+$ and M$^-$ are the values of the magnetization measured in an increasing and decreasing field respectively. Large samples can be oriented by magnetic texturing. Superconducting critical current properties of such bulk materials were studied [17-20]. The irreversible magnetization data also indicates steady improvement in the superconducting properties of the material with increasing annealing temperature up to 1055°C. The induced orientation within the sample is directly dependent on the magnetic field value as shown in Figure 2. In order to obtain a large degree of induced anisotropy, the magnetic field must be applied continuously in the appropriate temperature regime (from the maximal temperature down to a temperature below the liquidus line), during which crystals are still free to rotate. Steady increase in the induced anisotropy is observed when the magnetic field is increased.

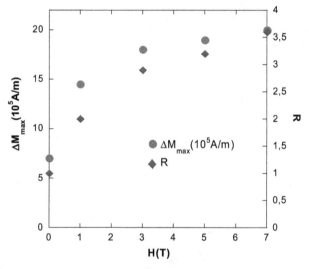

Figure 2. Variation in the magnitude of the magnetization hysteresis ΔM_{max}(●) and the ratio R(◊) with the strength of the annealing field at T_A=1055°C in $Y_1Ba_2Cu_3O_{7-\delta}$ bulk textured sample. The magnetic field was applied at all temperatures above 800°C [16].

It was also shown that the magnetic force exerted on a sample ($m\chi H_a \times dH_a/dz$) placed in a field gradient (dH_a/dz) allows the measurement of the magnetic susceptibility via the weight measurement on an electronic balance. The resulting curves provide information on the fusion, the solidification and the oxygen exchange. A typical curve presenting the magnetic susceptibility variations corrected from the oxygen weight change as a function of temperature is given in Figure 3.

When the sample is heated, the susceptibility first decreases following a Curie law behavior. When Y123 melts, between 1000 °C and 1040°C, the susceptibility increases because the melt is more paramagnetic than the phase. In the liquid state, above 1040°C, the oxygen stoichiometry changes and the mean ratio of copper valence is changed in the melt. The magnetic Cu^{2+} ions are progressively reduced in non-magnetic ions Cu^+ through the reaction

: $2CuO \rightarrow Cu_2O + 1/2O_2$ resulting in a rapid decrease of the susceptibility. When the reduction of the magnetic copper is taken into account via the mole fraction loss of oxygen, the susceptibility corrected obeys a classical Curie law as a function of temperature. Consequently above 1070°C, largely above the peritectic temperature, the nuclei composition changes since the melt is transformed and non-superconducting nuclei of secondary phases appear. It is the reason why the magnetic field is no more effective at very high temperature. The value of the supercooling can also be measured while cooling down.

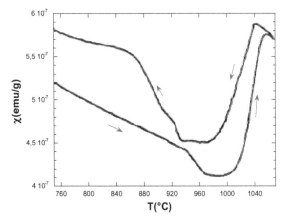

Figure 3. Magnetic susceptibility χ of Y$_1$Ba$_2$Cu$_3$O$_{7-\delta}$ at high temperature. Usually a change in the slope is attributed to a change of phase or to a change in composition.

Nowadays, top-seed melt texturing is mainly used to texture Y123 pellet where the crystallographic orientation is given by the crystal seed. The largest intrinsic nuclei which are involved in magnetic texturing are melted with an overheating temperature of about 1080 °C in order to be sure that crystallization is induced by the seed and not by intrinsic nuclei. In this technique, the growth of a main Y123 grain starts from a seed which has a higher peritectic decomposition temperature and crystallographic parameters close to Y-123. Pellets up to 10 cm of diameter were grown with this technique [21].

2.2. Case of Bi$_2$Sr$_2$CaCu$_2$O$_{8+x}$

2.2.1. Pellets of Bi$_2$Sr$_2$CaCu$_2$O$_{8+x}$

In Bi$_2$Sr$_2$CaCu$_2$O$_{8+x}$ (Bi2212), the achievement of the primary phase after melting is not easy. The system Bi$_2$O$_3$-SrO-CaO-CuO is composed of 4 elements. The phase equilibrium is complex and recombination of the primary phase from the melt is not always reversible when the composition in the liquid is too much changed. Small variations of composition or temperature induce large changes in the equilibrium and distributions of phases. Bi2212 phase can be in equilibrium with most of the system composites Bi$_2$O$_3$-SrO-CaO-CuO. Up to fifteen four-phase equilibriums were mentioned in the literature. The melting give rises to secondary non-superconducting phases that are not entirely consumed during solidification. There is a change in the liquid structure about 30 °C above the melting temperature of Bi2212 and Bi2223 in air as shown in Figure 4. [22,23]. The solidification of Bi2212 from the melt requires a precise control of the temperature and composition.

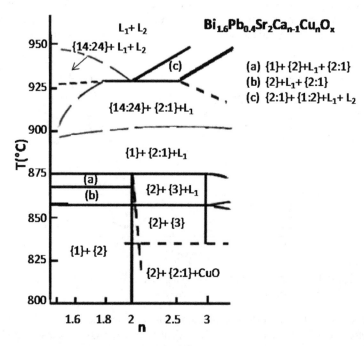

Figure 4. Part of the phase diagram of $Bi_{1.6}Pb_{0.4}Sr_2Ca_{n-1}Cu_nO_x$ as described in [23] for n varying between 1.5 and 3 and temperatures between 800 and 950°C.Phase {1} $Bi_2Sr_2Ca_0Cu_1O_x$ (n=1) ; phase {2} $Bi_2Sr_2Ca_1Cu_2O_x$ (n=2) ; phase {3} $Bi_2Sr_2Ca_2Cu_3O_x$ (n=3); phase {2:1} $(Sr,Ca)_2Cu_1$;phase {14:24} $(Sr,Ca)_{14}Cu_{24}$; phase {1:2} $(Sr,Ca)_1Cu_2$.

In BSSCO bulk compounds, several methods can be used and eventually mixed to induce alignment in the c-axis and overcome weak links. The magnetic field was successfully applied in this compound and important critical current densities were obtained [24-27]. A technique combining magnetic melt processing and hot forging was developed to improve the critical current densities in Bi2212 [28]. The magnetic field acts as an orienting tool while the pressing reinforces the orientation and the density. The addition of MgO in Bi2212 allows the material to keep high viscosity in the melt state at high temperature, making it compatible with a magnetic field texturing followed by hot forging. Moreover the MgO or Ag addition induces an improvement of superconducting properties.

Bi2212 powder added with 10 wt.% MgO is used as a starting powder. This mixture is cold pressed under an uniaxial pressure of 1 GPa in 20-mm diameter pellet with a thickness of 5 to 7 mm. The sample is melt-processed in a furnace inserted in the room temperature bore of a magnet. The melting temperature is determined by measuring the magnetic susceptibility as a function of temperature [6, 24,28].

The study of the magnetic susceptibility represented as a function of temperature in Figure 5 indicates the beginning of the melting process at 877°C and the end of the decomposition of the Bi2212 phases at 905°C [24]. Above 905°C, the liquid composition is stable and no phase is transformed so the magnetic susceptibility follows the classical Curie law and is proportional to 1/T. At 905°C, formation of the phases CuO and (Sr,Ca)O are observed. When the temperature is decreased, the susceptibility first increases before reaching a plateau. The beginning of the solidification can be seen when χ suddenly decreases. During cooling, the phase transformation at 740°C can be noted on the graph and corresponds to the solidification of the eutectic $Cu_2O/(Sr,Ca)_3Bi_2O_6$. In the inset figures, the susceptibility cycle was done for different maximal temperatures. One can observe that the susceptibility plateau is reached only for a maximal temperature of 892°C. Below 871°C, the susceptibility decreases during the cooling because the solidification process immediately takes place. The overheating temperature is not large enough in the interval 877-892°C. When the temperature is superior to 892°C, the susceptibility will reach the plateau before decreasing at the solidification. The liquidus temperature is equal to 892 °C [6].

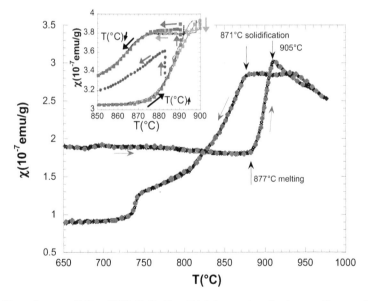

Figure 5. Magnetic susceptibility of $Bi_2Sr_2CaCu_2O_{8+x}$ at high temperature. Inset: magnetic susceptibility for different maximal temperatures. When the maximal temperature reached is inferior to 892°C, the magnetic susceptibility χ decreases during cooling. When the temperature reached is superior to 892°C, the susceptibility is first constant during cooling before decreasing [24].

The melting temperature being defined, the process can be optimized. In the process, the temperature can reach 1100°C and the vertical magnetic field used was 5.7 Tesla. The maximal processing temperature of 892°C was consequently defined as the optimum annealing temperature. Hot forging can be applied to bulk Bi2212/MgO magnetic melt processed. Hot Forging at 25 MPa during 2 hours at 880°C induces an increase in the orientation degree and of the density [28].

The magnetic melt processing performed 15°C above the onset of melting leads to the best orientation degree and thus to the best J_c. At this temperature, as was seen previously with YBaCuO the remaining magnetic susceptibility is still large enough to overcome the thermal disordering effect and there is enough liquid to allow free rotation of the crystallites. The anisotropy of susceptibility at 800 °C is equal to 4×10^{-8} emu/g [6]. For a magnetic field of 5.7T, and a temperature of 800°C, the crystals which are aligned in magnetic fields below 892 °C are larger than 10^{-21}m^3 as determined by (1). Above this temperature, the texture is gradually lost because of the transformation of Bi2212 nuclei in secondary phases leading to a difficulty in recombining the 2212 phase.

The results are consistent with the one of H. Maeda et al. where Bi2212 bulks and tapes were processed by the MMP in magnetic fields up to 15 T following the heat treatment shown in figure 6 [26]. A texture is also developed due to the anisotropy of magnetic susceptibility. The degree of texture and the anisotropy factor in magnetization increase almost linearly with the magnetic field strength during MMP. The anisotropy factor in magnetization reaches 3.2 and 6.5 at 13 T in Bi(Pb)2212 bulks and Ag-doped Bi2212 bulks respectively. For bulk materials, the doping of some fine particles, which induce melting and crystal nucleation, is required to achieve highly textured structures by MMP. The transport critical current density J_c and the transport critical current I_c of Bi2212 tapes increase with increasing Ha due to the texture development. These results indicate that MMP is effective to enhance the texture development and J_c values for Bi2212 bulks and tapes with thick cores, making it possible to fabricate tapes with high I_c.

These observations are also consistent with the fact that above the melting point, remaining nuclei subsist that tend to be aligned in the direction of the magnetic field when they are cooled in the window of solidification. However, leaving aside the difficulty to form different phases and to lose the initial composition when the overheating is too large, exceeding too much the melting temperature reduces the presence of the intrinsic nuclei which cannot act as growth crystals during solidification and the orientation and texture are progressively lost when the temperature is 30 °C above 892 °C.

2.2.2. Ag-sheated tapes of $Bi_2Sr_2CaCu_2O_{8+x}$

Magnetic texturing of Ag-sheathed tapes was studied in very high fields and lead to high critical current densities [27,29-32]. E. Flahaut demonstrated in his PhD the possibility to texture highly homogeneous Bi2212 tapes by moving the tape in the furnace and using a continuous melting process with large critical currents (J_c = 230kA/cm^2 at 4.2K, self field) [33].

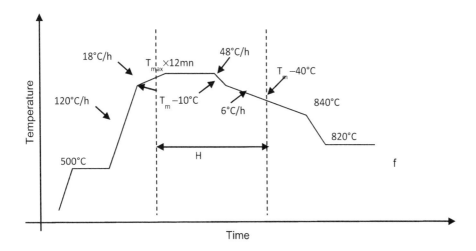

Figure 6. Generalized temperature profile for MMP Bi$_2$Sr$_2$CaCu$_2$O$_{8+x}$ tapes in magnetic field up to 15T [26]. T$_{max}$ is the maximal annealing temperature. The magnetic field is applied at T$_{max}$ and shut down at T$_{max}$ - 40°C. The atmosphere is flowing O$_2$.

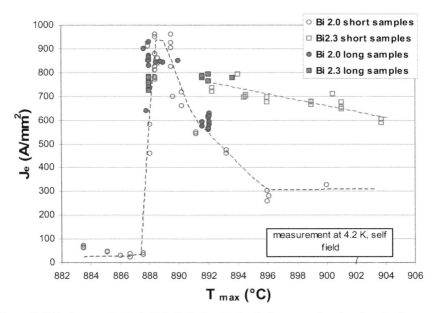

Figure 7. Critical current density in Bi$_2$Sr$_2$CaCu$_2$O$_{8+x}$ Ag-sheathed tapes as a function of maximal temperatures [34] for two different compositions. Pre-reacted Bi$_2$Sr$_2$CaCu$_2$O$_{8+x}$ powders are used.

It was shown that the critical current density can be improved with this method compared to a standard static heat treatment. The continuous process is applied to the fusion stage, the most sensitive treatment where the temperature must be carefully controlled, in order to improve the homogeneity of the sample. There are two reasons why a continuous process improves the critical current. Firstly, it allows the whole tape to be subjected to the same maximal temperature which is not the case in a furnace where the temperature must be perfectly homogeneous along the sample volume. Secondly, the texture can be improved by the thermal gradient that induces an orientation of the crystallites along the silver sheath of each filament. The engineering critical current density of Bi2212 tapes at 4.2K is plotted as a function of maximum annealing temperature used in the melting process in figure 7 [34]. The critical current is measured by a classical four point method. It is a good representation of the superconducting and crystallographic properties of the samples. Usually, a poor critical current is coming from a poor connectivity between the grains resulting from a lack of orientation and accumulation of secondary phases at the grain boundaries. The critical current density is maximal at the melting temperature and in a very narrow window of temperatures above the melting temperature. The Bi2212 nuclei acting in the magnetic texturing are also present in the narrow sheaths and induce the crystal growth. When the overheating temperature is too large, the critical current density progressively decreases for all the reasons mentioned above.

2.2.3. Magnetic field texturing of 2212 tapes using a continuous displacement in the furnace

Bi2212 multifilament tapes were processed by the well-established powder-in-tube (PIT) technique. The tape is composed of 78 filaments embedded in a Ag–Mg (0.2% Mg) sheath [35]. The tapes are different from the ones before. A dynamic heating process was inserted into a superconducting coil to enable a dynamic heat treatment under a magnetic field (5 T) in order to obtain homogeneous and high critical current densities (figure 8).

The thermal treatment consists in a melting step for 250 s and a slow temperature decrease (0.1 °C s^{-1}). The atmosphere is 40% flowing O_2. The maximum applied field is 5T. Critical current densities in Bi2212 tapes were measured in self field at 4.2K to compare the benefit of a magnetic field during the dynamic heat treatment. An increase of the critical current density is obtained under the application of a magnetic field at T= 894 °C as can be seen in figure 9. The processing speed is around 1cm/min. The results are reproducible and this gain in current densities is observed for temperature around 894°C. The results are consistent with the ones previously presented for Y123 and Bi2212 bulk samples. The rotation of the crystallites under a magnetic field is possible when the liquid at the interface of the grains is sufficient and its viscosity is high which implies a relatively high temperature of the melt. Above 896°C, the refractory and non-superconductive phases tend to grow rapidly, limiting the critical current density. The dynamic heat treatment under a magnetic field offers several benefits. Firstly, the temperature is highly homogeneous. Secondly, no long treatments are necessary and the texture can be obtained homogenously and rapidly. This is also due to the facts that the nuclei causing the growth of the right phase subsist in the melt and don't need the recombination in the solid state that requires long time.

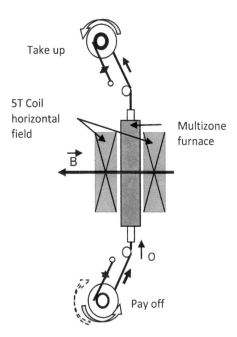

Figure 8. Dynamic heat process under a magnetic field. A multizone furnace is inserted in a superconducting coil reaching 5T. Flowing O$_2$, maximal temperature and processing speed are important parameters that can be controlled [35].

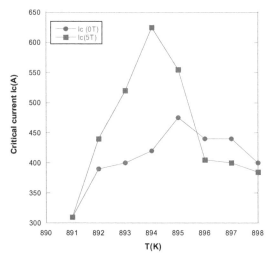

Figure 9. Critical current in Ag sheathed tapes of Bi$_2$Sr$_2$CaCu$_2$O$_{8+x}$ as a function of maximal annealing temperature measured at 4.2K with field and without field [35].

2.3. Case of Bi₂Sr₂Ca₂Cu₃O₈₊ₓ

2.3.1. Pellets of Bi₂Sr₂Ca₂Cu₃O₈₊ₓ

J. Noudem demonstrates the possibility to form the $Bi_2Sr_2Ca_2Cu_3O_{8+x}$ (Bi2223) phase after a liquid transformation above the melting temperature [36 - 38]. As seen previously in the Bi2212 compounds, the phase diagram is complex which requires a precise control of the temperature and composition [23].

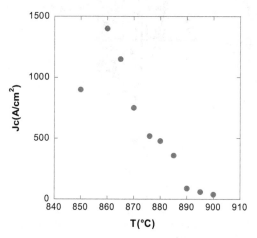

Figure 10. Critical current density in Bi2223 textured bulk compounds measured at 77K as a function of the maximal annealing temperature [38].

The texture along the c-axis is induced by means of a magnetic field. Pellets of Bi2223 were first uniaxialy pressed before annealing. The heating cycle took place in a vertical furnace placed in a superconducting magnet reaching 8T. The optimum maximum temperature lies within the range 855° and 900°C and is followed by a slow decrease in temperature of 2°C/h. A maximal value of critical current density of 1450 A/cm² was obtained for a temperature reaching 860°C during 1 hour. Below this value, the proportion of liquid phase is too small to allow rotation of the platelets under the influence of the magnetic field and the critical current consequently sharply decreases. Above this optimum processing temperature, the transformation of Bi2223 nuclei in secondary phases $(Ca/Sr)_{14}Cu_{24}O_y$, CuO and $(Ca/Sr)_2CuO_3$ prevent the recombination of Bi2223. Thus, the critical current density gradually decreased until zero around 900°C as shown in Figure 10 in agreement with the phase diagram of Figure 4.

2.3.2. Ag-sheated tapes of Bi₂Sr₂Ca₂Cu₃O₈₊ₓ

The possibility to obtain high critical current and a good homogeneity in Ag-sheathed Bi2223 tapes could follow accordingly to the previous results However, the complexity of the Bi,Pb(2223) phase diagram and the presence of numerous secondary phases coexisting with the Bi,Pb(2223) phase as shown in Figure 4 has led to a standard route of tapes preparation where precursor powders are calcined and pre-reacted before the cold deformation inside the Ag tubes. This route implies the use of Bi2212 phase as the predominant phase and a series of cold deformations (swaging, drawing and rolling) before heat-treated the Ag-sheathed tapes to achieve Bi,Pb(2223) phase. However, intrinsic limits to this route seem to come from the formation mechanisms where the transient liquid assisting the Bi,Pb(2223) formation is not stable and results in a poor structural homogeneity. E. Giannini et al showed that it is possible to form the Bi,Pb(2223) phase from a liquid close to the equilibrium conditions following a reversible melting of the Bi,Pb(2223) phase exactly in the same way as J. Noudem et al. did in bulk samples [39]. Then, Bi2223 intrinsic nuclei exist in the melt. This new route is extremely sensitive to the temperature profile and is dependent of the local Pb content [40]. This result confirms the possibility to texture Ag-sheathed Bi,Pb(2223) tapes by means of a scrolling continuous heat treatment under a magnetic field.

3. Magnetic texturing induced by intrinsic growth nuclei surviving above the melting temperature

3.1. Intrinsic nuclei surviving above the melting temperature

Undercooling temperatures of liquid alloys depend on the overheating rate above the melting temperature T_m of liquid alloys and elements [41-43]. The experiments presented in section 2 lead to the conclusion that intrinsic nuclei exist above T_m and are aligned in magnetic field during the growth time evolved in the solidification window between liquidus and solidus temperatures [7]. These results are in contradiction with the idea [8-10] that all crystals have to disappear at the melting temperature T_m by surface melting. The classical equation of Gibbs free energy change associated with crystal formation also predicts that the critical radius for crystal growth becomes infinite at T_m and all crystals are expected to melt [8-10]. It has been recently proposed to add an energy saving ε_v produced by the equalization of Fermi energies of out-of-equilibrium crystals and their melt in this equation [5,6].

Transformation of liquid-solid always induces changes of the conduction electron number per volume unit, and sometimes per atom. The equalization of Fermi energies of a spherical particle having a radius R smaller than a critical value R^*_{2ls} (θ) produces an unknown energy saving $-\varepsilon_v$ per volume unit. The ε_v value is equal to a fraction ε_{ls} of the fusion heat ΔH_m per molar volume V_m. This energy was included in the classical Gibbs free energy change $\Delta G_{1ls}(\theta,R)$ associated with crystal formation in metallic melts [5,6,10]. The modified free energy change is $\Delta G_{2ls}(R,\theta)$ given by (2) and $\Delta G_{1ls}(\theta,R)$ is obtained with $\varepsilon_{ls} = 0$:

$$\Delta G_{2ls}(R,\theta) = \frac{\Delta H_m}{V_m}(\theta - \varepsilon_{ls})4\pi\frac{R^3}{3} + 4\pi R^2 \frac{\Delta H_m}{V_m}(1+\varepsilon_{ls})(\frac{12k_B V_m \ln K_{ls}}{432\pi \times \Delta S_m})^{1/3} \qquad (2)$$

where ΔS_m is the molar fusion entropy, $\theta = (T-T_m)/T_m$ the reduced temperature, R the growth nucleus radius, $\varepsilon_{ls} \times \Delta H_m/V_m$ the energy saving ε_v, N_A the Avogadro number, k_B the Boltzmann constant. The lnK_{ls} given in (3) depends on the viscosity through the Vogel-Fulcher-Tammann (VFT) temperature T_{0m} of the glass-forming melt which corresponds to the free volume disappearance temperature [44,45]:

$$\ln(K_{lgs}) = \ln(\frac{A\eta_0}{\eta}) = (\ln A) \pm 2 - \frac{B}{(T-T_{0m})} \tag{3}$$

where lnA is equal to 87 ± 2, $B/(T_g-T_{0m}) = 36-39$, T_g being the vitreous transition temperature [46,47]. The critical radius and the critical energy barrier are respectively given versus temperature by (4) and (5) assuming that a minimum value of ε_v exists at each temperature $T \leq T_m$ and $\varepsilon_{ls} = 0$ for $R > R^*_{2ls}$:

$$R^*_{2ls}(\theta) = \frac{-2(1+\varepsilon_{ls})}{\theta-\varepsilon_{ls}}(\frac{V_m}{N_A})^{1/3}(\frac{12k_BN_A \ln K_{ls}}{432\pi \times \Delta S_m})^{1/3} \tag{4}$$

$$\frac{\Delta G^*_{2ls}}{k_BT} = \frac{12(1+\varepsilon_{ls})^3 \ln K_{ls}}{81(\theta-\varepsilon_{ls})^2(1+\theta)} \tag{5}$$

The critical radius at the melting temperature is no longer infinite at T_m ($\theta = 0$) because ε_{ls0} is not nil. Out-of-equilibrium crystals having a radius smaller than R^*_{2ls} are not melted because their surface energy barrier is too high. The thermal variation of ε_{ls} is an even function of θ given by (6):

$$\varepsilon_v = \varepsilon_{ls}\frac{\Delta H_m}{V_m} = \varepsilon_{ls0}(1-\frac{\theta^2}{\theta^2_{0m}})\frac{\Delta H_m}{V_m} \tag{6}$$

where θ_{0m} corresponds to the free-volume disappearance temperature T_{0m} [5]. The fusion entropy is always equal to the one of the bulk material regardless of the crystal radius R and of ε_{ls0} value as shown in (7) because $d\varepsilon_{ls}/dT = 0$ at the melting temperature T_m of the crystal :

$$\frac{3}{4\pi R^3}\left\{\frac{d[\Delta G_{2ls}(\theta)]}{dT}\right\}_{T=T_{mc}} = -\frac{\Delta S_m}{V_m} \tag{7}$$

All crystals outside a melt generally have a fusion temperature depending on their radius; they melt by surface melting. Here, they melt by homogeneous nucleation of liquid droplets. These properties are analogous to that of liquid droplets coated by solid layers that are known to survive above T_m [48,49]. The crystal stability could be enhanced by an interface thickness of several atom layers [50] or these entities could be super-clusters which could contain magic numbers of atoms [51,52]. They could melt by liquid homogeneous nucleation instead of surface melting. The presence of super-clusters in melts being able to act as growth nuclei was recently confirmed by simulation and observation in glass-forming melts and liquid elements [53-55].

The equalization of Fermi energies of out-of-equilibrium crystals and melts does not produce any charge screening induced by a transfer of electrons from the crystal to the melt as assumed in the past [5,6,47]. It is realized by a Laplace pressure p given by (8) [56,57] depending on the critical radius R^*_{2ls} and the surface energy σ defined by the coefficient of $4\pi R^2$ in (2):

$$ p = \frac{2 \times \sigma}{R^*_{2ls}} = \frac{\Delta H_m}{V_m} [\theta - \varepsilon_{ls}(\theta)] \tag{8} $$

The energy saving coefficient ε_{ls} and consequently p can be determined from the knowledge of the vitreous transition T^*_g in metallic and non-metallic glass-forming melts [56,57]. The liquid-glass transformation is accompanied by a weakening of the free volume disappearance temperatures from θ_{0m} to θ_{0g}, of energy saving coefficients at T_m from ε_{ls0} to ε_{lg0} and consequently of VFT temperatures from T_{0m} to T_{0g} [56]. The vitreous transition T^*_g occurs at a crystal homogeneous nucleation temperature T_{2lg} giving rise, in a first step, to vitreous clusters during a transient nucleation time equal to the glass relaxation time. The energy saving coefficients ε_{ls0} and ε_{lg0} are determined by θ^*_g respectively given by (9) and (10):

$$ \varepsilon_{lg0} = \varepsilon_{lg}(\theta = 0) = 1.5 * \theta^*_g + 2 \tag{9} $$

$$ \varepsilon_{ls0} = \varepsilon_{ls}(\theta = 0) = \theta^*_g + 2 \tag{10} $$

The free-volume disappearance reduced temperature θ_{0m} of the melt above T^*_g which is equal to its VFT temperature [45] is given by (11)]:

$$ \theta^2_{0m} = \frac{8}{9}\varepsilon_{ls0} - \frac{4}{9}\varepsilon^2_{ls0} \tag{11} $$

These formulae are applied to Bi2212 with $\theta^*_g = -0.42$ ($T^*_g = 676$ K); T^*_g is assumed to be 50 K below the first-crystallization temperature observed at 723 K [58]. Using (10), we find $\varepsilon_{ls0} = 1.58$ and $T_{0m} = 532$ K with a liquidus temperature of 1165 K [59]. The $\ln K_{ls}$ is equal to 76.4 with $\ln A = 85$ and $B/(T_m - T_{0m}) = 8.6$. The fusion heat ΔH_m is chosen to be equal to the mean value of 11 measurements depending on composition minus the heat produced by an oxygen loss of $1/3 \times$mole [22,25,60]. We obtain $\Delta H_m = 7100 - 2000 = 5100$ J/at.g, a molar volume $V_m = 144*10^{-6}$ m^3 and a fusion entropy 4.38 J/K/at.g. Another evaluation of ΔH_m leads to 4600 J using a measured oxygen loss of 2.1% of the sample mass in a magnetic field gradient maximum at the liquidus temperature T = 1165 K and a measured heat of 8000 J/at.g for a composition 2212 [60]. The critical radius R^*_{2ls} ($\theta = 0$) in the two cases are respectively equal to 8.94×10^{-10} and 9.27×10^{-10} m; these critical nuclei contain 188 and 209 atoms (Bi, Sr, Ca, Cu and O). Surviving nuclei have a smaller radius and contain less atoms. The energy saving coefficient ε_{nm0} of non-melted crystals has to be determined in order to explain why growth nuclei are of the order of 10^{24}-10^{25} per m^3 as shown by nano-crystallization of glass-forming melts and are able to induce solidification at T_m when the applied overheating rate remains weak [47,61].

3.2. Melting temperature of surviving intrinsic nuclei as a function of their radius

The radius R of liquid droplets which are created by homogeneous nucleation inside an out-of-equilibrium crystal of radius R_{nm} as a function of the overheating rate θ is calculated from the value of ΔG_{2sl} given by (12); (12) is obtained by replacing θ by $-\theta$ in (2) because the transformation occurs now from crystal to melt instead of melt to crystal:

$$\Delta G_{2sl}(R,\theta) = \frac{\Delta H_m}{V_m}(-\theta - \varepsilon_{nm0})4\pi\frac{R^3}{3} + 4\pi R^2\frac{\Delta H_m}{V_m}(1+\varepsilon_{nm0})(\frac{12k_B V_m \ln K_{ls}}{432\pi \times \Delta S_m})^{1/3} \qquad (12)$$

where ε_{ls} given by (6) is replaced by ε_{nm0} when $\theta \geq 0$ because the Fermi energy difference between crystal and melt is assumed to stay constant above T_m. A crystal melts at $T > T_m$ when its radius R_{nm} is smaller than the critical values 0.894 or 0.927 nm. The equation (13) of liquid homogeneous nucleation in crystals with the nucleation rate $J = (1/v.t_{sn})$ is used to determine $\Delta G_{2sl}/k_B T$ as a function of R_{nm} [7,10,47,62]:

$$\ln(J.v.t_{sn}) = \ln(K_{sl}.v.t_{sn}) - \frac{\Delta G_{sl}(R_{nm},\theta)}{k_B T} = 0 \qquad (13)$$

where v is the surviving crystal volume of radius R_{nm} and t_{sn} the steady-state nucleation time evolved at the overheating temperature T. The radius R_{nm} being smaller than 1 nanometer and t_{sn} chosen equal to 1800 s, $\ln(K_{sl}.v.t_{sn})$ is equal to 32-35 with $\ln K_{sl}$ being given by (3).

The equalization of Fermi energies in metallic melts is accomplished through Laplace pressure. We have imagined another way of equalizing the Fermi energies of nascent crystals and melt instead of applying a Laplace pressure [5,6,47]. Free electrons are virtually transferred from crystal to melt. The quantified energy savings of crystals having a radius R_{nm} equal or smaller than the critical value would depend on the number of transferred free electrons. A spherical attractive potential would be created and would bound these s-state conduction electrons. This assumption implies that the calculated energy saving ε_{nm0} at $T = T_m$ is quantified, depends on the crystal radius R_{nm} which corresponds to the first energy level of one s-electron moving in vacuum in the same spherical attractive potential. These quantified values of ε_{nm0} have been already used to predict the undercooling temperatures of gold and other liquid elements in perfect agreement with experiments [63,64]. As already shown, the attractive potential $-U_0$ defined by (14) is a good approximation for $n \times \Delta z \gg 1$, $n = 4\pi N_A R^3/3V_m$ being the atom number per spherical crystal of radius R_{nm} and Δz the fraction of electron per atom which would be transferred in vacuum from crystal to melt, e the electron charge, $\varepsilon_{nm0} \times \Delta H_m$ the quantified electrostatic energy saving per mole at T_m, ε_0 the vacuum permittivity:

$$U_0 = \frac{n\Delta z\, e^2}{8\pi\varepsilon_0 R_{nm}} \geq E_q = \frac{n\varepsilon_{nm0}\Delta H_m}{N_A} \qquad (14)$$

The potential energy U$_0$ is nearly equal to the quantified energy E$_q$ with ε_{ls0} = 1.58 at T = T$_m$ and Δz = 0.1073 with a critical radius of 0.894 nm, ΔH_m = 5100 J/atom.g, using (14). We find Δz = 0.0994 with a critical radius of 0.927 nm and ΔH_m = 4600 J. Schrödinger's equation (15) is written with wave functions ψ only depending on the distance r from the potential centre for a s-state electron [65]:

$$\frac{1}{r}\frac{d^2}{dr^2}(r\psi) + k^2\psi = 0, \text{ where } k = \frac{1}{\hbar}[2m(U_0 - E_q)]^{1/2} \tag{15}$$

The quantified solutions

$$E_q = \frac{n\varepsilon_{nm0}\Delta H_m}{N_A}$$

are given by the k value and by (16) as a function of the potential U$_0$ associated with a crystal radius R = R$_{nm}$, n varying with the cube of R$_{nm}$:

$$\frac{\sin kR_{nm}}{kR_{nm}} = \frac{\hbar}{\sqrt{2mR_{nm}^2 U_0}}, \text{ where } U_0 = \frac{n\Delta z e^2}{8\pi\varepsilon_0 R_{nm}} \tag{16}$$

The quantified values of E$_q$ at T$_m$ and consequently of ε_{nm0} are calculated as a function of R$_{nm}$ from the knowledge of U$_0$ using a constant Δz. The free-energy change given by (12) is plotted versus R$_{nm}$ in Figure 11 and corresponds to a liquid droplet formation of radius R$_{nm}$ at various overheating rates θ. Crystals are melted when (13) is respected with $\Delta G_{2sl}/k_B T$ equal or smaller than ln(K$_{ls}$.v.t$_{sn}$). All crystals having a radius larger than the critical value 0.894 nm and smaller than 0.36 nm are melted at T$_m$ while those with 0.36 < R$_{nm}$< 0.894 nm are not melted. It is predicted that, for the two ΔH_m values, they would disappear at temperatures respecting θ > 0.93 assuming that all atoms are equivalent. It is known that a consistent overheating has to be applied to glass-forming melts in order to eliminate a premature crystallization during cooling [41]. A glass state is obtained when Bi-2212 melts are quenched from 1473-1523 K (0.264 < θ < 0.307) after 1800 s evolved at this temperature [60,61]. Many nuclei survive after an overheating at θ = 0.264. Bi2201 grains of diameter 10 nm are obtained after annealing the quenched undercooled state at 773 K during 600 s [61]. Consequently, the nuclei number is still larger than 10^{24}/m^3 in spite of an overheating at 1473 K (θ = 0.264) in agreement with the curves $\Delta G_{2sl}/k_B T$ which are larger than ln(K$_{sl}$.v.t$_{sn}$) $\cong$ 20 with t = 1800 s and v < 1 nm^3.

3.3. Crystallization of melts induced by intrinsic nuclei at T ≤ T$_m$

These intrinsic nuclei act as growth nuclei during a solidification process. The first-crystallization will occur when (17) will be respected at the temperature T:

$$\ln(J.v.t_{sn}) = \ln(K_{sl}.v_s.t_{sn}) - \left[\frac{\Delta G_{sl}^*(\theta)}{k_B T} - \frac{\Delta G_{nm}(\theta)}{k_B T}\right] = 0 \tag{17}$$

where $\Delta G^*_{2ls}/k_BT$ is given by (5) and $\Delta G_{nm}/k_BT$ is the contribution of unmelted crystal of radius R_{nm} to the reduction of $\Delta G^*_{2ls}/k_BT$ given by (2).

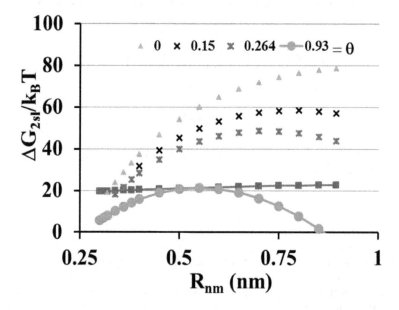

Figure 11. Values of ΔG_{2sl} given by (12) divided by k_BT and calculated with a radius R equal to an intrinsic nucleus radius R_{nm} smaller than the critical radius equal to 0.894 nanometer are plotted versus R_{nm} for $\theta = (T-T_m)/T_m = 0, 0.1, 0.15, 0.264$ and 0.93. The straight line nearly parallel to the R_{nm} axis represents $\ln(K_{ls}.v.t_{sn})$ with v equal to the volume of a sphere having a radius R_{nm}. All nuclei with $\Delta G_{2sl}/k_BT > \ln(K_{ls}.v.t_{sn})$ are not melted. The lower radius below which all nuclei melt slowly increases with θ. All nuclei would disappear after 1800 s at T = 2248 K ($\theta = 0.93$).

The $\ln K_{ls}$ given by (3) is equal to 76.4 at T_m while the volume sample v_s is assumed to be equal to 10^{-6} and 10^{-24} m³ respectively leading to $\ln(K_{ls}.v_s.t_{sn}) = 70.1$ and $\ln(K_{ls}.v_s.t) = 28.6$ with $t_{sn} = 1800$ s. In Figure 12, $(\Delta G^*_{2ls}/k_BT-\Delta G_{nm}/k_BT) = \Delta G_{eff}/k_BT$, $\ln(K_{ls}.v_s.t_{sn}) = 70.1$ and 28.6 are plotted as a function of R_{nm} at T = T_m. Nuclei with $0.45 < R_{nm} < 0.894$ nm could act as growth nuclei at T_m. Nuclei with R_{nm} smaller than 0.45 nm contain about one or two layers of atoms surrounding a centre atom. All metallic glass-forming melts contain the same type of clusters that govern their time-temperature-transformation (TTT) diagram above T_g [47]. The Bi2212 TTT diagram has to exist even if it is not known. In contradiction with our model of identical atoms, intrinsic nuclei having a radius R_{nm} respecting $0.45 < R_{nm} < 0.894$ nm have to be melted after an overheating up to 1473 K ($\theta = 0.264$) because a weak cooling rate leads to a vitreous state in this type of glass-forming melt after applying an overheating rate $\theta = 0.264$ [61].

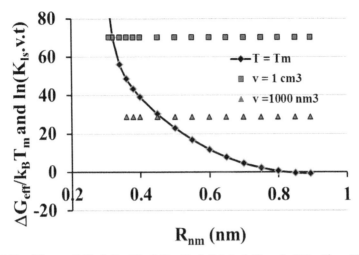

Figure 12. The difference ($\Delta G^*_{2ls}/k_BT_m - \Delta G_{nm}/k_BT_m = \Delta G_{eff}/k_BT_m$), the $\ln(K_{ls}.v_s.t) = 70.1$ with $v = 10^{-6}$ m^3, $K_{ls} = 76.4$, $t_{sn} = 1800$ s, and $\ln(K_{ls}.v.t) = 28.6$ with $K_{ls} = 76.4$, $v = 10^{-24}$ m^3 and $t_{sn} = 1800$ s are represented as a function of the surviving nucleus radius R_{nm}. All nuclei with $0.45 < R_{nm} < 0.894$ nm can grow at $T = T_m$ in a sample volume of 10^{-24} m^3 using our model of identical atoms.

The calculation of the first-crystallization nucleation total time t has to take into account not only the steady-state nucleation time t_{sn} but also the time-lag τ_{ns} in transient nucleation [44] with:

$$t = t_{sn} + \frac{\pi^2}{6}\tau_{ns}, \text{ when } t >> \frac{\pi^2}{6}\tau_{ns} \tag{18}$$

$$J\tau^{ns} = \frac{a_0^*N}{2\pi\Gamma}\exp(-\frac{\Delta G_{eff}^*}{k_BT}), \text{ where } \Gamma = (\frac{1}{3\pi k_BT}\frac{\Delta G_{2ls}^*}{j_c^2})^{1/2}, \quad \tau^{ns} = \frac{a_0^*N}{2\pi K_{ls}\Gamma} \text{ and } j_c = \frac{32\pi\times\alpha_{2ls}^3}{3(\theta-\varepsilon_{nm})^3} \tag{19}$$

The time lag τ_{ns} is proportional to the viscosity η while the steady-state nucleation rate J is proportional to η^{-1}; the $J\tau_{ns}$ in (19) is not dependent on η [44]; Γ is the Zeldovitch factor, $a^*_0 = \pi^2/6$, N the atom number per volume unit, j_c the atom number in the critical nucleus; K_{ls} is defined by (3). Equation (17) is also applied when t_{sn} is small compared with the time lag τ_{ns}. A very small value of t_{sn}/τ_{ns} undervalues the time t by a factor 2 or 3 and has a negligible effect in a logarithmic scale. The first crystallization occurs when $J = (v.t_{sn})^{-1}$ in the presence of one intrinsic growth nuclei in the sample volume v. The critical energy barrier is obtained using (5) and the values of ε_{nm} as a function of θ^2; t_{sn} is deduced from (20) with $\ln(K_{ls})$ given by (3).

Assuming that the intrinsic nuclei are very numerous regardless the sample volume v, the steady-state nucleation rate J per volume unit and per second is given by:

$$\ln(J.v.t_{sn}) = \ln(K_{ls}v.t_{sn}) - \frac{\Delta G_{eff}^*}{k_BT}, \text{ where } \frac{\Delta G_{eff}^*}{k_BT} = \frac{\Delta G_{2ls}^*}{k_BT} - \frac{\Delta G_{nm}}{k_BT} \tag{20}$$

where t_{sn} is the steady-state nucleation time and ΔG_{nm} the free-energy change associated with the previous solidification of non-melted crystals [8-10,44,62]].

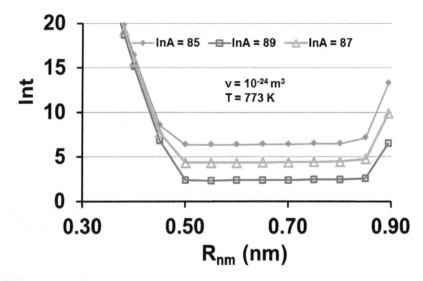

Figure 13. The logarithm of the nucleation full time t is plotted as a function of the nucleus radius R_{nm} using three values of $\ln A$, $\Delta H_m = 5100$ J. and the corresponding values of $\ln K_{ls}$ at $T_x = 773$ K. The shortest nucleation times are equal to 584, 79 and 11 s as compared to an experimental value of 600 s. This nucleation time is dominated by the transient time τ_{ns} which becomes much larger at $T_g = 676$ K and equal to 2×10^{18} s. These results lead to $(T_x - T_g) \cong 97$ K measured with a heating rate of 10K/mn. Similar curves are obtained using $\Delta H_m = 4600$ J and are not represented.

The steady–state homogeneous-nucleation time t_{sn} ($\Delta G_{nm} = 0$) necessary to observe a first crystallization would be equal to 10^{24} seconds with a sample volume $v = 10^{-24}$ m³. The energy barrier ΔG_{nm} of intrinsic nuclei is calculated using (2); the radius R_{nm} of intrinsic nuclei is temperature-independent down to the growth temperature; ε_{nm} (θ) is assumed to vary as ε_{ls} (θ) given by (3) with ε_{ls0} replaced by ε_{nm0} which is calculated using (14-16). A unique nucleus in a volume $v_s = 10^{-24}$ m³ still acts as growth nucleus and produces a nano-crystallization process of the undercooled melt as shown in Figure 13.

The Bi2212 melt contains more than 10^{24} nuclei per m³ in full agreement with nanocrystallization experiments [61]. The values of $\ln A$ are equal to 87 ± 2 as already shown for glass-forming melts having about the same T_g/T_m and V_m. The total nucleation time t is calculated by adding t_{sn} and τ_{ns} with (18). The shortest nucleation time is equal to 584, 79 and 11 s for $\ln A = 85$, 87 and 89 respectively. The nanocrystallization has been observed at $T = 773$ K after 600 s in perfect agreement with the curve $\ln A = 85$. The model developed above has already predicted the time-temperature transformation diagram of several glass-forming melts [47].

Surviving intrinsic nuclei have been called non-melted crystals. These clusters exist in the melt above T$_m$ and they have a stability only expected in super-clusters which could correspond to a complete filling of electron shells [51,52]. This assumption might be correct if the structural change of surviving crystals in less-dense superclusters occurs without complementary Gibbs free energy. The Gibbs free energy change (12) induced by a volume increase would remain constant because it only depends on $R/(V_m/N_A)^{1/3}$. An increase of the cluster molar volume would be followed by an increase of R without change of ε_{nm0} and U_0 in (15), (16) and (8). Then, surviving intrinsic nuclei could be super-clusters already observed in liquid aluminium by high energy X-rays [55]. We develop here the idea that the super-clusters are solid residues instead of belonging to the melt structure.

The model described here also predicts the undercooling temperatures of liquid elements [64]. It only requires the knowledge of the melting temperature of each nucleus as a function of their radius, the fusion heat of a bulk multicomponent material, its molar volume, its melting temperature and its vitreous transition temperature. Adjustable parameters are not required. There is an important limitation to its use because all atoms are considered as being identical. Thus, it cannot describe the behaviour at high temperatures of clusters which have a composition strongly different from that of the melt. We did not find any publication about the existence of a vitreous state of Y123. This could be due to a much smaller viscosity of this melt. Nevertheless, this material can be magnetically textured due to the presence of intrinsic clusters above T$_m$. Its unknown energy saving coefficient ε_{ls0} could be smaller than 1 and could not be determined using equations only governing fragile glass-forming melts. The growth of nuclei occurs inside the temperature interval between liquidus and solidus. The magnetic field can give a direction to the free crystals when their size depending on their magnetic anisotropy is sufficiently large; the processing time below the liquidus temperature has to be adapted to their growth velocity.

4. Conclusion

Magnetic texturing experiments using pre-reacted powders of Bi2212, Bi2223 or Y123 show that crystal alignment in magnetic field is successful when the annealing temperature is slightly superior to the critical temperature T$_m$. This temperature corresponds to the liquidus temperature T$_m$ above which there is no solid structure except isolated growth nuclei imbedded in the melt having radii smaller than a critical value. The growth around nuclei occurs in a window of solidification extending from T$_m$ and leads to crystals which become free to align in the magnetic field when their anisotropy energy becomes larger than k_BT. The thermal cycle has to use an annealing temperature a little larger than T$_m$ in order to be sure that crystallisation will start from surviving nuclei of superconducting phases and not from polycrystalline entities. The overheating temperature has to be limited to a few degrees Celsius up to a few tens of degrees above T$_m$ depending on the amplitude of the magnetic field. This will insure that the number of nuclei per volume unit remains important and that the composition of nuclei is not progressively transformed into non-superconducting secondary phase nuclei accompanying the oxygen loss.

Short and long lengths of Bi2212 conductors obtained from pre-reacted powders of Bi2212 were textured using the same thermal cycle with a maximal annealing temperature a little larger than T_m and a relatively small annealing time. This thermal cycle is to be compared to the usual long annealing time imposed at lower temperatures in order to form the superconducting phase from non-reacted precursors. This process is successful even without magnetic field because the melt already contains surviving nuclei of this phase acting as growth nuclei inside the silver-sheathed filaments having a section sufficiently small to accommodate the platelets obtained after growth. The alignment of these platelets and consequently the superconducting critical current are improved by using a dynamic process reproducing the thermal cycle presented above where the wire moves through the magnetic field.

We have used a model which predicts glass-forming melt properties without using adjustable parameters and only knowing the fusion heat of a bulk multicomponent material, its molar volume, its melting temperature and its vitreous transition temperature T_g. The Bi2212 and Bi2223 ceramics are known to lead to a vitreous state after quenching. The knowledge of T_g leads to the homogeneous nucleation temperature of crystal and to the energy saving associated with the complementary Laplace pressure acting on a critical radius crystal which is associated with the Fermi energies equalization of this crystal and melt at all temperatures T larger than T_g. The melting temperature of each nucleus above T_m depends on its radius which is smaller than its critical value at T_m. A simple method of quantification is used to determine the complementary Laplace pressure depending on the crystal radius. This energy saving of surviving crystals is calculated assuming that the equalization of Fermi energies between crystal and melt virtually occurs by free electron transfer from crystal to melt. A bound state of a s-state electron would be produced in the electrostatic spherical potential of radius R_{nm}; the energy saving depending on R_{nm} would be equal to its first energy level. This energy is constant above T_m at constant radius and decreases with $\theta^2 = (T-T_m)^2/T_m^2$ below T_m. The transient and steady-state nucleation times of the Bi2212 nano-crystallized state are predicted in agreement with the observed values. There is a limitation to the model since we consider all atoms to be identical. Consequently, it cannot describe the behaviour at too high temperatures of clusters which have a composition strongly different from that of the melt. The model predicts that the highest melting temperature of surviving crystals occurs at $\theta = 0.93$. This value is too high because the vitreous state is obtained by quenching the melt from $\theta \cong 0.3$. In addition, Bi2212 and, Bi2223 nuclei are rapidly transformed in secondary phase clusters which govern the nano-crystallization of the melt. We did not find any publication about the existence of a vitreous state of YBCO. This could be due to a much smaller viscosity of this melt. Nevertheless, this material can be magnetically textured due to the presence of intrinsic clusters above T_m. Its unknown energy saving coefficient ε_{lso} could be smaller than 1 and could not be determined using equations only governing fragile glass-forming melts.

Author details

Laureline Porcar, Patricia de Rango, Daniel Bourgault and Robert Tournier
Institut Néel/CRETA/CNRS/ University Joseph Fourier/Grenoble/France

Acknowledgments

The authors thank the Institute of Physics Publishing, IOP publishing, Taylor and Francis Group Journals, Elsevier for giving the authorization to reproduce Figure 1 and 2 as published in [16], Figure 4 as published in [23], Figure 6 as published in [26], Figure 7 as published in [34], Figure 8 and 9 as published in [35] and Figure 10 as published in [38].

5. References

[1] Beaugnon E, Bourgault D, Braithwaite D, de Rango P, Perrier de la Bâthie R, Sulpice et al. Material processing in high static magnetic field: a review of an experimental study of levitation, phase separation, convection and texturation. J de Physique,1993; 3:399-421.

[2] Yamaguchi M, Tanimoto Y. Magneto-Science: Magnetic Field Effects on Materials: Fundamentals and Applications. Ed: Springer Series in Materials Science vol 89, Tokyo: Kodansha. 2006.

[3] Tournier R. Method for preparing an oriented and textured magnetic material, U.S.A. Patent, 1992; Dec. 1, N° 5,168,096.

[4] Tournier R, Crystal making method. U.S.A. patent, 1993; june 8, N° 5,217,944.

[5] Tournier, RF. Presence of intrinsic growth nuclei in overheated and undercooled liquid elements. Physica B, 2007; 392:79-91.

[6] Tournier RF. Tiny crystals surviving above the melting temperature and acting as growth nuclei of the high –Tc superconductor microstructure. Mat. Sc. Forum, 2007; 546-549:1827-40.

[7] Tournier R F, Beaugnon E. Texturing by cooling a metallic melt in a magnetic field. Sci. Technol. Adv. Mater. 2009; 10:014501 (10pp), available from: URL: http://stacks.iop.org/1468-6996/10/014501.

[8] Turnbull D. Kinetics of solidification of supercooled liquid mercury droplets. J. Chem. Phys.1952; 20: 411-24.

[9] Kelton K F. Crystal nucleation in liquids and glasses, Solid State Phys. 1995; 45:75 -177.

[10] Vinet B, Magnusson L, Fredriksson H, Desré PJ. Correlations between surface and interface energies with respect to crystal nucleation. J. Coll.Interf. Sc. 2002; 255:363-374.

[11] Tong HY, Shi FG. Abrupt discontinuous relationships between supercooling and melt overheating. Appl. Phys. Lett. 1997; 70:841-43

[12] Mikelson AE, Karklin YK. Control of crystallization processes by means of magnetic fields. J. Cryst. Growth, 1981; 52:524-29

[13] Ferreira JM, Maple M B, Zhou H, Hake RR, Lee BW, Seaman CL, Kuric M V et al. Magnetic field alignment of high-Tc Superconductors RBa$_2$Cu$_3$O$_{7-\delta}$. Appl. Phys. A , 1988; 47:105-10.

[14] Lees MR, de Rango P, Bourgault D, Barbut JM, Braithwaite D, Lejay P et al. Bulk textured rare-earth-Ba$_2$Cu$_3$O$_{7-\delta}$ prepared by solidification in a magnetic field. Supercond. Sci. Technol.1992; 5:362-7.

[15] De Rango P, Lees M, Lejay P, Sulpice A, Tournier R., Ingold M et al, Texturing of magnetic materials at high temperature by solidification in a magnetic field, Nature, 1991; 349:770-1.

[16] Lees M, Bourgault D, de Rango P, Lejay P, Sulpice A, Tournier R. A study of the use of a magnetic field to control the microstructure of the high-temperature superconducting oxide $YBa_2Cu_3O_{7-\delta}$ Phil. Mag. B, 1992; 65:1395-404.

[17] Bourgault D, de Rango P, Barbut JM, Braithwaite D, Lees MR, Lejay P et al. Magnetically melt-textured YBa_2 $Cu_3O_{7-\delta}$ Physica C, 1992; 194, pp. 171-6

[18] Barbut JM, Barrault M, Boileau F, Ingold F, Bourgault D, de Rango P, Tournier R. D.C. transport current in excess 1000A in a superconducting material textured by a zone melting set-up in a magnetic field. Appl. Superc.1993; 1:345-8.

[19] Barbut JM, Bourgault D, Schopohl N, Sulpice A, Tournier R. Scaling laws Jc (H/Hc3(⊚)) and intrinsic surface superconductivity in YBa_2 $Cu_3O_{7-\epsilon}$. Physica C, 1994; 235-240:2995-6.

[20] Porcar L, Bourgault D, Barbut JM, Barrault M, Germi P, Tournier R. High transport currents of melt textured YBCO up to 6000 A, Physica C, 1997; 275:293-8.

[21] Tournier R, Beaugnon E, Belmont O, Chaud X, Bourgault D, Isfort D et al. Processing of large $YBa_2Cu_3O_{7-x}$ single domains for current-limiting applications. Supercond. Sci. Technol. 2000; 13:886-95.

[22] Strobel P, Korczak W, Fournier T. A thermal analysis study of the system Bi-Sr-Ca-Cu-O at variable oxygen pressure. Phys. C,1989; 161:167-74.

[23] Strobel P, Toledano JC, Morin D, Schneck J, Vacquier G, Monnereau O et al. Phase diagram of the system $Bi_{1.6}Pb_{0.4}Sr_2CuO_6$-$CaCuO_2$ between 825°C and 1100°C. Physica C, 1992; 201:27-42.

[24] Pavard S, Villard C, Bourgault D, Tournier R. Effect of adding MgO to bulk Bi2212 melt textured in a high magnetic field., Supercond. Sci. Technol. 1998; 11:1359-66.

[25] Chen WP, Maeda H, Kakimoto K, Zhang PX, Watanabe K, Motokawa M. Processing of Ag-doped Bi2212 bulks in high magnetic fields: a strong correlation between degree of texture and field strength, Physica. C, 1999; 320:96-100.

[26] Maeda H, Chen WP, Inaba T, Sato M, Watanabe K, Motokawa M. Texture development in Bi-based superconductors grown in high magnetic fields and its effect on transformation of Bi(Pb)2212 to Bi(Pb)2223, Physica C, 2001; 354:338-41.

[27] Maeda H, Sastry PVPSS, Trociewitz UP, Schwartz J, Ohya K, Sato et al. Effect of magnetic field strength in melt-processing on texture development and critical current density of Bi-oxide superconductors. Physica C, 2003; 386:115–121.

[28] Tournier R, Pavard S, Bourgault D, Villard C. Bulk Bi2212 texturing by solidification in a high magnetic field and hot forging, Int. Symp. on Superc. XII, ISS 99, Morioka, Japan,1999; pp. 527-9.

[29] Liu HB, Ferreira PJ, Vander Sande JB. Processing Bi-2212/Ag thick films under a high magnetic field: on the Bi-2212/Ag interface effect. Physica C, 1998; 303:161-168.

[30] Chen WP, Maeda H, Watanabe K, Motokawa M, Kitaguchi H, Kumakura H. Microstructures and properties of Bi 2212 Ag tapes grown in high magnetic fields. Physica C, 1999; 324:172-6.

[31] Maeda H, Ohya K, Sato M, Chen WP, Watanabe K, Motokawa M, et al. Microstructure and critical current density of Bi2212 tapes grown by magnetic melt-processing, Physica C, 2002; 382:33-7.
[32] Liu X, Schwartz J. On the influence of magnetic field processing on the texture, phase assemblage and properties of low aspect ratio Bi2Sr2CaCu2Ox/AgMg wire. Sci. Technol. Adv. Mater., 2009, 10, 014605 (11pp), available from: URL: http://stacks.iop.org/STAM/10/014605.
[33] Flahaut E, Bourgault D, Bruzek CE, Rikel MO, Herrmann P, Soubeyroux, J.-L.,et al. R. Dynamic heat treatment of BSCCO-2212 tapes with homogeneous properties and high critical current density. IEEE-Trans.-Appl.-Supercond. June 2003; 13(2) pt. 3: 3034-7.
[34] Bruzek C-E, Lallouet N, Flahaut E, Bourgault D, Saugrain J-M, Allais A et al. High performance Bi2212/Ag tape produced at Nexans. IOP Conf. Series, 2003; 181:2260-7.
[35] Villaume A, Bourgault D, Porcar L, Girard A, Bruzek CE, Sibeud PF. Misalignment angles reduction in Bi2212 multifilaments melted by dynamic heat treatment under a magnetic field. Supercond. Sci. Technol. 2007; 20:691-6.
[36] Noudem JG, Beille J, Bourgault D, Sulpice A, Tournier R. Combined effect of melting under magnetic field and hot pressing to texture superconducting ceramics Bi-Pb-Sr-Ca-Cu-O (2223). Physica C,1994; 235-240:3401-2
[37] Noudem JG, Beille J, Bourgault D, Beaugnon E, Chateigner D, Germi P et al. Magnetic melt texturation combined with hot pressing applied to superconducting ceramics Bi-Pb-Sr-Ca-Cu-O (2223). Supercond. Sc. Technol.1995; 8:558-63
[38] Noudem J, Beille J, Bourgault D, Chateigner D, Tournier R. Bulk Bi-Pb-Sr-Ca-Cu-O (2223) ceramics by solidification in a magnetic field. Physica C, 1996; 264:325-30.
[39] Giannini E, Savysyuk I, Garnier V, Passerini R, Toulemonde P, Flükiger R. Reversible melting and equilibrium phase formation of (Bi,Pb)$_2$Sr$_2$Ca$_2$Cu$_3$O$_{10+\delta}$. Supercond. Sci. Technol. 2002, 15:1577-86.
[40] Li JY, Soubeyroux J-L, Zheng HL, Li CS, Lu YF, Porcar L, Tournier R et al. Phase evolution during the melting and recrystallization of ceramic core in the (Bi,Pb)-2223 tape. Physica C, 2006; 450:56–60.
[41] Hays CC, Johnson WL. Undercooling of bulk metallic glasses processed by electrostatic levitation. J. Non-Cryst. Solids, 1999; 250-252:596-600.
[42] Fan C, Inoue A. Influence of the liquid states on the crystallization process of nanocrystal-forming Zr-Cu-Pd-Al metallic glasses. Appl. Phys.Lett. 1999; 75:3644-6.
[43] Rudolph P, Koh HJ, Shäfer N, Fukuda T. The crystal perfection depends on the superheating of the mother phase too - experimental facts and speculations on the "melt structure" of semiconductor compounds. J. Cryst. Growth, 1996; 166:578-82.
[44] Gutzow I, Schmelzer. The vitreous state. Ed : Springer, 1995.
[45] Doolittle AK. Studies in Newtonian Flow. II. The dependence of the viscosity of Liquids on Free-Space. J. Appl. Phys. 1951; 22:1471-5.
[46] Angell C. A. Formation of glasses from liquids and biopolymers, Science, 1995; 267, pp 1924-1935.
[47] Tournier RF. Crystal growth nucleation and Fermi energy equalization of intrinsic spherical nuclei in glass-forming melts. Sci. Technol. Adv. Mater. 2009; 10:014607 (12pp), available from: URL: http://stacks.iop.org/1468-6996/10/01450.

[48] Lu K, Li Y. Homogeneous nucleation catastrophe as a kinetic stability limit for superheated crystal. Phys. Rev. Lett. 1998; 80:4474-7.

[49] Jiang Q, Zhang Z, Li JC. Melting thermodynamics of nanocrystals embedded in a matrix. Acta Mater. 2000; 48:4791-5.

[50] Reedjik MF, Arsic J, Hollander F FA, De Vries SA, Vlieg E. Liquid order at the interface of KDP crystals with water: evidence for ice-like layers. Phys. Rev. Lett. 2003; 90:066103

[51] De Heer WA. The physics of simple metal clusters: experimental aspects and simple model, Rev. Mod. Phys. 1993; 65:611-76.

[52] Kulmin VI, Tytik DL, Belashchenko DK, Sirenko A N. Structure of silver clusters with magic numbers of atoms by data of molecular dynamics. Colloid J. 2008; 70:284-96.

[53] Liu R-S, Liu H-R, Dong K-J, Hou Z-Y, Tian Z-A, Peng A-B et al. Simulation study of size distributions and magic number sequences of clusters during the solidification process in liquid metal Na. J. Non-Cryst.Sol. 2009; 355:541–47.

[54] Almyras GA, Lekka CE, Mattern N, Evangelakis GA. On the microstructure of the $Cu_{65}Zr_{35}$ and $Cu_{35}Zr_{65}$ metallic glasses. Scripta Mater. 1996, 62:33-36.

[55] Mauro NA, Bendert J C, Vogt AJ, Gewin JM, Kelton KF. High energy x-ray scattering studies of the local order in liquid Al. J. Chem. Phys. 2011, 135 (4):044502-1-6.

[56] Tournier RF. Thermodynamics of vitreous transition. Revue de Métallurgie, 2012; 109:27-33.

[57] Tournier RF. Thermodynamic and kinetic origins of the vitreous transition. Intermetallics, 2012; doi:10.1016/j.intermet.2012.03.024.

[58] Ramanathan S, Li Z, Ravi-Chandar K. On the growth of BSCCO whiskers. Phys. C, 1997; 289:192-98.

[59] Tournier RF. Thermodynamic origin of the vitreous transition, Materials, 2011, 4:869-92. available from URL: http://www.mdpi.com/journal/materials

[60] Mayoral MC, Andres JM, Bona M T, Angurel L A, Natividad E. Application to the laser floating zone: preparation of high temperature BSSCO superconductors by DSC. Thermochimica Acta, 2004; 409:157-164.

[61] Müller C, Majewski P, Thurn G, Aldinger F. Processing effects on mechanical and superconducting properties of Bi 2201 and Bi 2212 glass ceramics. Phys. C, 1997; 275:337-45.

[62] Turnbull D, Fisher JC. Rate of nucleation in condensed sytems. J. Chem. Phys. 1948; 17:71-3.

[63] Tournier RF. Expected properties of gold melt containing intrinsic nuclei, Proc. 6th Intern. Conf. Electrom. Process. Mat., Dresden, Ed: Forschungszentrum Dresden-Rossendorf, Germany,2009, pp. 304-307.

[64] Tournier RF. Undercooling versus Overheating of Liquid Elements Containing Intrinsic Nuclei: Application to Magnetic Texturing. Ppt presentation at MAP4 in Atlanta, 2010, available from URL: http://creta.cnrs.fr.

[65] Landau L, Lifchitz E. Mécanique Quantique. Editions MIR, Moscou,1966 ; pp.135-136.

Chapter 8

Pseudogap and Local Pairs
in High-T_c Superconductors

Andrei L. Solovjov

Additional information is available at the end of the chapter

1. Introduction

The discovery of superconductivity in copper oxides with an active CuO_2 plane [1] at temperatures of the order of 100 K is undoubtedly one of the most important achievements of the modern solid state physics. However, even more than twenty six years later since the discovery the physics of the electronic processes and interactions in high-temperature superconductors (HTS's) and, in particular, the superconducting (SC) pairing mechanism, resulting in such high T_c's, where T_c is the superconducting transition (critical) temperature, still remain controversial [2]. This state of affairs is due to the extreme complexity of the electronic configuration of HTS's, where quasi-two-dimensionality is combined with strong charge and spin correlations [3, 4, 6–12].

Gradually it became clear that the physics of superconductivity in HTS's can be understood, first and foremost, by studying their properties in the normal state, which are well known to be very peculiar [4, 6–12]. It is believed at present that the HTS's possess at least five specific properties [2, 7, 8, 11–13]. First of all it is the high T_c itself which is of the order of 91 K in optimally doped (OD, oxygen index $(7 - \delta) \approx 6.93$) $YBa_2Cu_3O_{7-\delta}$ (YBCO, or Y123), of the order of 115 K in OD $Bi_2Sr_2Ca_2Cu_3O_{8+\delta}$ (Bi2223) and in corresponding Tl2223 [8, 9, 12], and arises up to $T_c \approx 135K$ in OD $HgBa_2Ca_2Cu_3O_{8+\delta}$ (Hg1223) cuprates [14]. The next and the most intrigueing property is a pseudogap (PG) observed mostly in underdoped cuprates below any representative temperature $T^* \gg T_c$ [2, 8]. As T decreases below T^*, these HTS's develop into the PG state which is characterized by many unusual features [2, 8, 12, 13, 15, 16]. The other property is the strong electron correlations observed in the underdoped cuprates too [3, 5, 12, 16]. However, existence of the such correlations in, e.g., FeAs-based superconductors still remains controversial [17–21]. The next property is pronounced anisotropy [6–9, 11, 12] observed both in cuprates [2, 9, 12] and FeAs-based superconductors (see Refs. [17–19] and references therein). As a result, the inplane resistivity, $\rho_{ab}(T)$ is much smaller than $\rho_c(T)$, and the coherence length in the ab plane, $\xi_{ab}(T)$, is about ten times of the coherence length along the c-axis, $\xi_c(T)$. The last but not least property is a reduced density of charge carriers n_f. n_f is zero in the antiferromagnetic (AFM) parent state of HTS's and gradually increases with doping [2, 8, 9, 11, 12]. But even in an optimally doped YBCO it is an order of magnitude less than in conventional superconductors [2, 8, 9, 11, 12, 15]. There is growing evidence that just the reduced density of charge carriers may be a key feature to account for all other properties of HTS's [2, 6, 7, 13, 22–26].

The Chapter addresses the problem of the PG which is believed to appear most likely due to the ability of a part of conduction electrons to form paired fermions (so-called local pairs) in a high-T_c superconductor at $T \leq T^*$ [6, 13, 22–27]

2. Theoretical background

There are two different approaches to the question of the mechanisms for SC pairing of charge carriers in cuprates and therefore the physical nature of the PG [2, 27]. In the first approach, pairing of charge carriers in HTS's is of a predominantly electronic character, and the influence of phonons is inessential [3–5, 28, 29]. In the second approach, pairing in HTS's can be explained within the framework of the Bardeen-Cooper Shieffer (BCS) theory, if its conclusions are extended to the case of strong coupling [12, 30–32]. However, it gradually became clear that aside from the well-known electron-phonon mechanism of superconductivity, due to the inter-electronic attraction by means of phonon exchange [33, 34], other mechanisms associated with the inter-electronic Coulomb interaction can also exist in HTS's [3–8, 10, 12, 35, 36]. That is why it is not surprising that the systems of quasiparticle electronic excitations, where factors other than phonons and excitons resulting in inter-electronic attraction and pairing are considered, are studied in a considerable number of theoretical investigations of HTS's. Some examples are charge-density waves [7, 37, 38], spin fluctuations [3, 11, 39–41], "spin bag" formation [42, 43], and the specific nature of the band structure - "nesting" [44]. The main distinguishing feature of these investigations compared to the conventional superconductors is the more detailed study of the models based on the existence of strong interelectronic repulsion in the Hubbard model which can result in anisotropic d-pairing [35, 45]. Attempts have also been made to construct anisotropic models of high-temperature superconducting (HTS) systems with different mechanisms of interelectronic attraction [12, 46, 47]. Unfortunately, the consensus between both approaches has not been found. As a result, up to now there is still no completed fundamental theory to describe high-T_c superconductivity as a whole and to clarify finally the PG phenomenon [2].

2.1. Pseudogap in HTS's

Gradually it became evident that a high critical temperature is by no means the only property that distinguishes HTS's from conventional low-temperature superconductors. Another, property of cuprates, which attracts much attention, is the PG state [2, 8, 15, 26, 36]. All experiments convincingly show [8] that as the charge-carrier density decreases relative to its value in OD samples, a completely unusual state, where the properties of the normal and superconducting phases appear together [48], arises in HTS systems in a sizable temperature interval above T_c [8, 15, 26, 27, 49–54]. From the very discovery of the PG state some authors called this state a "pseudogap phase". However, for example, as Abrikosov notes [55], this state actually cannot be interpreted as some new phase state of matter, since the PG is not separated from the normal state by a phase transition. At the same time it can be said that HTS system undergoes a crossover at $T \leq T^*$ [56]. Below $T^* >> T_c$, for reasons which have still not been finally established, the density of quasiparticle states at the Fermi level starts to decrease [57–59]. That is why the phenomenon has been named a "pseudogap"

The number of papers devoted to the problem of the pseudogap in HTS's is extraordinarily large (see Refs. [6–8, 12, 15, 84] and [40, 49, 50, 53, 54, 56, 59] and references therein) and new papers are constantly appearing [38, 41, 47, 60–63]. It seems to be reasonable as it is completely clear that a correct understanding of this phenomenon can also provide an answer to the question of the nature of high-temperature superconductivity as a whole. Among many other papers it is worth to mention the most radical model as for the nature of high-temperature superconductivity and a PG in cuprates. It is the Resonating Valence Bonds (RVB) model proposed by Anderson [64, 65], which describes a spin liquid of singlet electronic pairs. In this model, largely relies on the results obtained using one-dimensional models of interacting electrons, the low-temperature behavior of electrons differs sharply from the standard behavior in ordinary three-dimensional (3D) systems. An electron possessing charge and spin is no longer a well-defined excitation. So-called charge and spin separation occurs. It is supposed that spin is transferred by an uncharged fermion, called a spinon and charge - a spinless excitation - by a holon. In the RVB model both types of excitations - spinons and holons - contribute to the resistivity. However, the holon contribution is considered to be determining, while spinons, which are effectively coupled with a magnetic field **H**, must determine the temperature dependence of the Hall effect. Even though the RVB model led to a series of successes [65, 66], it is difficult to think up the physics behind the processes which could lead to the charge and spin separation especially in quasi-two-dimensional systems, which cuprate HTS's are. Nevertheless, the RVB model contains at lest one rational idea, namely, it is supposed that two kinds of quasi-particles with different properties have to exist in the high-temperature superconducting (HTS) system at T below T^*. In the RVB model such particles are spinons and holons.

However, even though researchers have made great efforts in this direction, the physics of the PG phenomenon is still not entirely understood (see Ref. [2] and references therein). That is why, we have eventually to propose our own Local Pair (LP) model [13, 27] developed to study a pseudogap $\Delta^*(T)$ in high-temperature superconductors and based on analysis of the excess conductivity derived from resistivity experiments. We share the idea of the RVB model as for existence of two kinds of quasi-particles with different properties in HTS's below T^*. But in our LP model these are normal electrons and local pairs, respectively. I will frame our discussion in terms of the local pairs, and try to show that this approach allows us to get a set of reasonable and self-consistent results and clarify many of the above questions.

2.2. The main considerations as for local pair existence in HTS's

There are several considerations leading to the understanding of the possibility of paired fermions existence in HTS's at temperatures well above T_c which I am going to discuss now. It is well known, that a pseudogap in HTS's is manifested in resistivity measurements as a downturn of the longitudinal resistivity $\rho_{xx}(T)$ at $T \leq T^*$ from its linear behaviour above T^* (Fig.1). This results in the excess conductivity $\sigma'(T) = \sigma(T) - \sigma_N(T)$, which can be written as

$$\sigma'(T) = [\rho_N(T) - \rho(T)]/[\rho(T)\rho_N(T)]. \tag{1}$$

Here $\rho(T) = \rho_{xx}(T)$ and $\rho_N(T) = \alpha T + b$ determines the resistivity of a sample in the normal state extrapolated toward low temperatures.

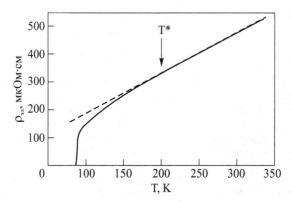

Figure 1. Resistivity ρ versus temperature T ($\bullet$) for YBCO film F1 (Table I); dashed line represents $\rho_N(T)$.

This way of determining $\rho_N(T)$, which is widely used for calculation $\sigma'(T)$ in HTS's [2], has found validation in the Nearly Antiferromagnetic Fermi Liquid (NAFL) model [11]. The question of whether the appearance of excess conductivity $\sigma'(T)$ in cuprates can be wholly attributed to fluctuating Cooper pairing or whether there are other physical mechanisms responsible for the decrease of $\rho_{xx}(T)$ at $T < T^*$ is one of the central questions in the modern physics of HTS's. To clarify the issue it seems reasonable to probe the PG by studying the fluctuation (induced) conductivity (FLC). The study of FLC provides the relatively easy but rather effective method which directly examines the possibility of paired fermions arising at temperatures preceding their transition into the SC state [67, 68].

Both Aslamasov-Larkin (AL) [69] and Maki-Thompson (MT) [70, 71] coventional FLC theories have been modified for the HTS's by Hikami and Larkin (HL) [72]. In the absence of a magnetic field the AL contribution to the FLC is given by the expression

$$\sigma'_{AL} = \frac{e^2}{16\hbar d}(1 + 2\alpha)^{-1/2}\varepsilon^{-1}, \tag{2}$$

Correspondingly, the HL theory gives for the MT fluctuation contribution the equation

$$\sigma'_{MT} = \frac{e^2}{8d\hbar}\frac{1}{1 - \alpha/\delta}\ln\left((\delta/\alpha)\frac{1 + \alpha + \sqrt{1 + 2\alpha}}{1 + \delta + \sqrt{1 + 2\delta}}\right)\varepsilon^{-1}. \tag{3}$$

In both equations

$$\alpha = 2\,[\xi_c^2(T)\,/\,d^2] = 2\,[\xi_c(0)\,/\,d]^2\,\varepsilon^{-1} \tag{4}$$

is the coupling parameter, $d \simeq 11.7$ Å in YBCO, is the distance between conducting layers,

$$\delta = 1,203\frac{l}{\xi_{ab}}\frac{16}{\pi\hbar}\left[\frac{\xi_c(0)}{d}\right]^2 k_B\,T\,\tau_\phi \tag{5}$$

is the pair-breaking parameter, and ξ_c is the coherence length along the c-axis, i.e. perpendicular to the CuO_2 conducting planes. The factor $\beta = 1.203(l\,/\,\xi_{ab})$, where l is the electron mean-free path, and ξ_{ab} is the coherence length in the ab plane, takes account of the approach to the clean limit introduced in the Bierei, Maki, and Thompson (BMT) theory [73]. Correspondingly,

$$\varepsilon = ln(T/T_c^{m}f) \approx (T - T_c^{mf}) / T_c^{mf} \qquad (6)$$

is the reduced temperature in HTS's. Here $T_c^{mf} > T_c$ is the critical temperature in the mean-field approximation, which separates the FLC region from the region of critical fluctuations or fluctuations of the order parameter Δ directly near T_c, neglected in the Ginzburg-Landau (GL) theory [74, 75]. Hence it is evident that the correct determination of T_c^{mf} is decisive in FLC calculations.

Equation (3) actually reproduces the result of the Lawrence-Doniach (LD) model [76], which examines the behavior of the FLC in layered superconductors, which cuprates and FeAs-based superconductors actually are. In the LD model, and hence in the HL theory, it is proposed that a Josephson interaction is present between the conducting layers. This occurs in the 3D temperature region, i.e., near T_c, where $\xi_c(T) > d$. Thus, according to the HL theory the AL fluctuation contribution predominates near T_c. Correspondingly, the MT mechanism predominates for $k(T - T_c^{mf}) >> \hbar/\tau_\phi$, where two-particle tunnelling between conducting layers is impossible, since $\xi_c(T) < d$ (2D fluctuation region) [77]. Thus, the HL theory predicts the alteration of the electronic dimensionality of the HTS sample leading to a 2D-3D crossover, as T approaches T_c. Simultaneously the physical mechanism of superconducting fluctuations changes too resulting in MT-LD crossover. In accordance with the HL theory, the 2D-3D crossover occurs at

$$T_0 = T_c\{1 + 2[\xi_c(0)/d]^2\} \qquad (7)$$

where it is assumed that $\alpha=1/2$, i.e.

$$\xi_c(0) = (d/2)\sqrt{\varepsilon_0}. \qquad (8)$$

Thus, now $\xi_c(0)$ can be determined, since ε_0 is a measured reduced crossover temperature. Correspondingly, the MT-LD crossover occurs at a temperature at which $\delta \simeq \alpha$ [72], which gives

$$\varepsilon_0 = (\pi \hbar)/[1.203(l/\xi_{ab})(8 k_B T \tau_\phi)] \qquad (9)$$

In accordance with our results [67], it should be the same temperature T_0. It makes it possible to determine τ_ϕ which is the phase relaxation time (lifetime) of fluctuating pairs. The evaluation of τ_ϕ in comparison with transport relaxation time τ of charge carriers measured by electrical conductivity, is a principal contribution to understanding the physics of transport properties. Whether $\tau_\phi > \tau$ or $\tau_\phi \approx \tau$ is important in view of the controversy over the Fermi-liquid or non-Fermi-liquid nature of the electronic state in HTS's [9, 64, 77]. Thus, the study of FLC can yield information about the scattering and fluctuating pairing mechanisms in HTS's as T draws near T_c.

As it was shown in our study of FLC [2, 67, 68], for optimally doped YBCO the interval $T_c < T < T_{c0} = (110 \pm 5) K$ is precisely that temperature region in which the temperature dependence of the resistivity, and consequently of the excess conductivity, is governed by the superconducting fluctuations leading to the onset of fluctuation conductivity which is described by the conventional fluctuation theories [69–72], as mentioned above. It means that fluctuating Cooper pairs have to exist in cooprates up to very high temperatures, namely, up to ~ 130 K in YBCO [27, 67, 68, 78] and up to $\sim (140 - 150)$ K in Bi compounds [16, 52, 53].

The conclusion has subsequently been shown to be consistent with results of several other research groups which will be briefly discussed now. 1. Kawabata et al. [78] has made a number of small ($D \sim 3\mu km$) holes in the slightly underdoped YBCO film by means of photolithography and then applied a magnetic field. Expected magnetic flux quantization was observed up to $T_{pair} \sim 130\,K$. The important point here is that period of oscillations corresponds to the charge of $Q = 2e$ evidently suggesting the electronic pairing in this temperature range. 2. In tunneling experiments by, e.g., Renner et al. [79], the peculiarities of measured differential conductivity dI/dV observed in the SC part of the PG in Bi2212 compounds [13] were found to persist up to temperatures well above T_c, and disappeared only at $T = T_{pair} \approx 140\,K$. But the wide maximum corresponding to the non-SC part of the PG was observed up to $T^* \approx 210\,K$. 3. It has subsequently been shown to be consistent with results of other groups dealing with the tunneling measurements [50–54]. Thus, in tunneling experiments by Yamada et al. performed on Bi compounds too [52], the temperature dependencies of the SC gap and PG, equal to the positions of the tunnel conductivity peaks, similar to that obtained in Ref [79], were studied for the three Bi2223 samples with different doping level. The noticeable increase of the PG in temperature interval from T_c up to $T_{pair} \approx 150\,K$ (SC part of the PG), similar to that obtained in our experiments with YBCO films [27], was found for all three samples [52]. However, in accordance with the LP model [13, 27], the peaks, which have the SC nature, as well as corresponding PG values, are smeared out above T_{pair}, suggesting expected transition into non-SC part of the PG.

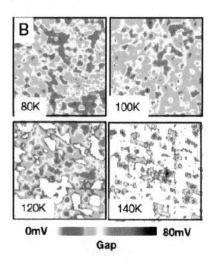

Figure 2. STM image of Bi2212 with T_c =93 K at different temperatures [[16]].

4. Eventually, Yazdani [16] was able to get the direct image of the local pair SC clusters in Bi compounds up to approximately $140\,K$ using the novel STM technique (Fig. 2). As it is clearly seen in the figure obtained for an optimally doped Bi2212 sample (T_c = 93 K), the state of the system at 100 K is almost the same as that below T_c. At 120 K the picture evidently changes, but the local pairs still form the SC cluster which determines the collective (superconducting-like) behavior of the system in this SC part of the PG. Even at 140 K the

SC nuclei are distinctly seen in the figure. But there are no closed clusters now. As a result, no collective behavior of the system is observed above $T_{pair} \geq 140\,K$, as it was shown in the tunneling experiments [52]. In accordance with the LP model, it is a non-SC part of a PG above T_{pair} [13]. Taking all these experimental results into account, the existence of paired fermions (local pairs) in the SC part of the PG, i.e, in the temperature interval from T_c up to T_{pair}, is believed to be well established now [13]. Additionally, T_{pair} is found to amount to $\simeq 130, 140$ and $150\,K$ for Y123, Bi2212, and Bi2223, respectively [13, 80].

2.2.1. Properties of the systems with low and reduced current-carrier density

Before to proceed with a question what would happen with the local pairs in the non-SC part of the PG above T_{pair} [13], let us have a look once again at the resistivity curve (Fig.1) obtained for our slightly underdoped YBCO film (sample F1, Table I). Three representative temperatures, at which $\rho(T)$ noticeably changes it slop, are distinctly seen on the plot. The first temperature is $T^* = 203\,K$ at which the local pairs are believed to appear [2, 27]. The second one is $T_{on} \approx 89,4\,K$ corresponding to the onset of the superconducting transition. The last one is $T_c = 87.4\,K$ at which the local pairs have to condense [6, 23, 24]. One basic question is whether any particularities affect the slop of the experimental curve around $T_{pair} \sim 130\,K$. The answer is completely negative. Indeed, the resistivity smoothly evolves with temperature and shows no peculiarities up to T^*. The fact suggests that nothing happens with the pairs at T_{pair}. Thus, one may draw a conclusion that if there are paired fermions in the sample below T_{pair} they also have to exist at $T > T_{pair}$, i.e., up to the very T^*. The point of view where the appearance of a PG in HTS's is due to the paired fermions formation at $T_c < T < T^*$ gradually gaining predominance [15, 27, 47, 81–84]. The possibility of the long-lived pair states formation in HTS's in the PG temperature range was justified theoretically in Refs. [26, 85, 86]. Nevertheless, the question of whether or not paired fermions can form in HTS's in the whole PG temperature range still remains very controversial. Indeed, it seems unlikely that conventional fluctuating Cooper pairs [33] are formed at temperatures $T^* > 200\,K$ [13, 27] especially considering the fact that the coherence length in HTS's is extremely short ($\xi_{ab}(T \leq T^*) \simeq (10 \div 15)$ Å) [2, 6, 8, 11, 12, 15].

We have, however, to keep in mind that we are dealing with the systems with low and reduced charge-carrier density n_f, as mentioned above. It has been shown theoretically [6, 23–26, 85] that such systems acquire some unusual properties compared to conventional superconductors. In conventional superconductors it is assumed that the chemical potential $\mu = \varepsilon_F$, where $\varepsilon_F = E_F$ is the Fermi energy, and their relation actually depends on nothing. In the systems with low and reduced n_f the chemical potential μ becomes a function of n_f, T and the energy of a bound state of two fermions, $\varepsilon_b = -(m\,\xi_b^2)^{-1}$ [23–26]. Here ξ_b is the scattering length in the s channel, and m is the mass of fermions with a quadratic dispersion law $\varepsilon(k) \sim k^2$ [23–25]. In the case of HTS's it is believed that $\xi_b(T)$ equals to the coherence length of a superconductor in the ab plane, $\xi_{ab}(T)$, and $m = m^*$ which is an effective mass of quasi-particles [11, 47]. ($m^* \sim 4.7m_0$ in nearly optimally doped (OD) YBCO [9, 11, 67, 68]. Thus, the ε_b becomes an important physical parameter of a Fermi liquid and determines a quantitative criterion for dense ($\varepsilon_F \gg |\varepsilon_b|$) or dilute ($\varepsilon_F \ll |\varepsilon_b|$) Fermi liquid. Accordingly, $k_F\varepsilon_b \gg 1$ and $\mu = \varepsilon_F$ in the first case, and $k_F\varepsilon_b \ll 1$ and $\mu = -|\varepsilon_b|/2$ ($\neq \varepsilon_F$) in the second one which correspond to the strong coupling [61, 84, 86]. Consequently, in the strong-coupling limit μ is to be equal to approximately $\varepsilon_b/2$. It should be also noted, that in this case the paired fermions have to appear in the form of so-called strongly bound bosons (SBB) which satisfy

Bose-Einstein condensation (BEC) theory [6, 23–26, 84–89]. In accordance with the theory, the SBB are extremely short but very tightly coupled pairs. As a result, the SBB have to be local (i.e. not interacting with one another) objects since the pair size is much less than the distance between the pairs. Besides, they cannot be destroyed by thermal fluctuations, and consequently may form at very high temperatures.

It is clear that some other parameters, including the mean-field critical temperature T_c^{mf} and temperature dependence of the SC order parameter $\Delta(T)$ have to change too. Analysis shows that in conventional superconductors with a high fermion density $T_c^{mf} \approx T_c$, i.e. it is identical to the BCS theory value [6, 23, 24, 33]. Moreover, $T_c^{mf} << \varepsilon_F$ in this case. For the systems with low density $T_c^{mf} \sim |\varepsilon_b|$, whence $T_c^{mf} >> \varepsilon_F$ [6, 24, 25]. The latter relation means that in this case T_c^{mf} characterizes not the condensation temperature T_c but rather the temperature at which the fermions start to bind into pairs, i.e., T^*. An equation for $\Delta(T)$ in a form convenient for comparing with experiment was obtained in Ref [87]. In this case the temperature dependence $\Delta(T)$ was calculated on the basis of the crossover from BCS to BEC limit for different values of the parameter $\mu/\Delta(0) = x_0$, where $\Delta(0)$ is the value of the SC order parameter at T=0:

$$\Delta(T) = \Delta(0) - \left((8\sqrt{\pi}) \sqrt{-x_0(\Delta(0)/T)^{3/2}} \right) exp\left[-(\mu^2 + \Delta^2(0))^{1/2}/T \right] \qquad (10)$$

Equation (2) determines how the character of $\Delta(T)$ changes when the parameter x_0 changes from 10 (BCS limit) to -10 (BEC limit) (Fig. 12). As it is shown in Ref. [88] the character of the pseudogap temperature dependence $\Delta^*(T)$ in HTS's has to change in the same manner as the charge-carrier density decreases.

2.2.2. The model of the local pairs (Local Pair model)

For obvious reasons, the question of which density should be regarded as low or high has not been posed for ordinary metals, and for a long time the question of a supposed BCS-BEC transition with decreasing n_f [74] was only of theoretical interest. The situation changed dramatically after the discovery of HTS's [1], where the charge-carrier density n_f is much lower than in conventional superconductors [8, 9, 11, 12, 15, 27], as discussed above. It means that in HTS's the mentioned above strongly bound bosons have to exist. Besides, the coherence length in the ab plane, $\xi_{ab}(T)$ is extremely short in HTS's, especially at high temperatures [2, 6, 8, 11, 12, 15] ($\xi_{ab}(T \leq T^*) \simeq$ 13Å in YBCO [67, 68]). It leads to the very strong bonding energy ε_b in the pair [23–25] which is an additional requirement for the formation of the SBB [6, 24–26]. Taking all above considerations into account, one may draw a conclusion that at high temperatures ($T \leq T^*$) the local pairs in HTS's, which are believed to generate a pseudogap [2, 6, 13, 26, 89], have to be in the form of the SBB [2, 6, 13, 26, 27, 47, 81, 85–89]. This is the first basic assumption of the LP model [2, 27]. This condition is realized just in the underdoped cuprates (see Ref. [2] and references therein) and new FeAs-based superconductors [17, 20]. But, strictly speaking, the presence or absence of a PG in FeAs-based HTS's still remain controversial [17, 21].

This assumption is supported by the fact that in accordance with the theory [6, 23, 24, 26] fermions start to bind into pairs at T^*, whereas the local pairs (or SBB) formed in the process may condense only for $T_c << T^*$, which at first glance seems to be in complete agreement with experimental observations [2, 12, 80, 90, 91]. But, the non-interacting SBB cannot be condensed at all [6, 23, 24, 85, 86], and this is a point. Eventually, just the value of $\xi_{ab}(T) = \xi_{ab}(0) (T/T_c - 1)^{-1/2}$ will determine the system behavior below T^* [2, 6, 24, 25, 86–89]. As temperature lowers, $\xi_{ab}(T)$ has to noticeably increase whereas the bonding energy ε_b in the pair has to decrease. As a result, the paired fermions have to change their state from the SBB into fluctuating Cooper pairs which behave in a good many ways like those of conventional superconductor [2, 6, 13, 26, 85, 89]. It is just that we call the local pairs. Thus, with decrease of temperature there must be a transition from BEC to BCS state, which is a consequence of a very short ξ_{ab} at high temperatures and its noticeable temperature dependence. *The possibility of a such transition is the main assumption of the LP model [2, 13, 27].* Precisely how this happens is one of the challenging questions in strongly correlated electron systems. Nevertheless, the transition was predicted theoretically in Refs. [23–25, 89] and approved in our experimental studies [2, 27].

Within the LP model a new approach to the analysis of the FLC and PG in HTS's was developed [2, 13, 27, 67]. First, it was convincingly shown that FLC measured for all without exception HTS's always demonstrates a transition from 2D ($\xi_c(T) < d$) into 3D ($\xi_c(T) > d$) state, as T draws near T_c. The result is most likely a consequence of Gaussian fluctuations of the order parameter in 2D metals [6, 23–25, 84], which HTS compounds with pronounced quasi-two-dimensional anisotropy of conducting properties actually are [2, 6, 9]. The Gaussian fluctuations were found to prevent any phase coherency organization in 2D compounds. As a result, the critical temperature of an ideal 2D metal is found to be zero (Mermin-Wagner-Hoenberg theorem), and a finite value is obtained only when three-dimensional effects are taken into account [6, 23, 24, 85, 87]. That is why, the FLC in the 3D state is always extrapolated by the standard 3D equation of the AL theory, which determines the FLC in any 3D system:

$$\sigma'_{AL3D} = \frac{e^2}{32\,\hbar\,\xi_c(0)}\varepsilon^{-1/2}, \tag{11}$$

This means that the conventional 3D FLC is realized in HTS's as $T \to T_c$ [67, 77]. Above the crossover temperature T_0 (Eq. (7)) the FLC in well-structured YBCO films was found to be of the MT FLC type [2, 67], in a good agreement with the HL theory [72]. The LD model was found to describe the experimental FLC only in the case of HTS compounds with pronounced structural defects [92, 93]. Therefore, we denote the observed 2D-3D crossover also as MT-AL [2, 67], unlike the MT-LD one predicted by the HL theory. It is clear on physical grounds that with increasing temperature the 3D fluctuation regime will persist until $\xi_c > d$ [77]. Thus, in this case the crossover should occur at $\xi_c \cong d$, i.e., at

$$\xi_c(0) = d \sqrt{\varepsilon_0}, \tag{12}$$

which is larger by a factor of two than is predicted by the LD and HL theories. $\xi_c(0)$ is one of the important parameters of the PG analysis.

Second, observation of the 2D-3D (MT-AL) crossover allows us to determine ε_0 quite accurately and, using Eq.(12), to obtain reliable values of $\xi_c(0)$ [2, 67, 68]. However, τ_ϕ [(see Eq.(9)] still remains unknown, since neither l nor $\xi_{ab}(0)$ is measured experimentally in a study of FLC. To find τ_ϕ we proceed as follows: we denote

$$\beta = [1.203\,(l/\xi_{ab})]; \tag{13}$$

we assume as before that $\tau_\phi(T) \propto 1/T$ [11, 67], and for our subsequent estimate of $\tau_\phi(100\,K)$ we assume that $\tau_\phi T$ =const. Finally, equation (9) can be rewritten as

$$\tau_\phi \beta T = (\pi \hbar)/(8 k_B T \varepsilon_0) = A \varepsilon_0, \qquad (14)$$

where $A = (\pi \hbar)/(8 k_B) = 2.988 \times 10^{-12}$. Now the parameter $\tau_\phi(100\,k) \beta$ is also clearly determined by the measured value of ε_0 and eventually enables us to determine $\tau_\phi(100\,k)$ [2, 67, 68].

Third, now as a PG analysis is concern. It is clear, to get information about the PG we need an equation which describes the whole experimental curve from T_c up to T^* and contains PG in the explicit form. Besides, the dynamics of pair-creation and pair-breaking above T_c must be taken into account [2, 32, 45, 72, 84]. However, the conventional fluctuation theories [72] well fit the experiment up to approximately 110 K only, whereas $T^* \simeq 200\,K$ even in slightly underdoped cooprates [2, 27, 67], as discussed above. Unfortunately, so far there is no completed fundamental theory to describe the high-T_c superconductivity as a whole and in particular the pseudogap phenomenon. For lack of the theory, such equation for $\sigma'(\varepsilon)$ has been proposed in Ref. [27] with respect to the local pairs:

$$\sigma'(\varepsilon) = \frac{e^2 A_4 \left(1 - \frac{T}{T^*}\right) \left(exp\left(-\frac{\Delta^*}{T}\right)\right)}{(16 \hbar \, \xi_c(0) \sqrt{2 \varepsilon_{c0}^*} \, \sinh(2\varepsilon / \varepsilon_{c0}^*)}, \qquad (15)$$

where ε is a reduced temperature given by Eq. (6), and $T_c^{mf} > T_c$ is the mean-field critical temperature, as discussed above. Besides, the dynamics of pair-creation $((1 - T/T^*))$ and pair-breaking $(exp(-\Delta^*/T))$ above T_c have been taken into account in order to correctly describe the experiment [2, 13, 27]. Here A_4 is a numerical factor which has the meaning of the C-factor in the FLC theory [2, 92]. All other parameters, including the coherence length along the c-axis, $\xi_c(0)$, and the theoretical parameter ε_{c0}^* [2, 27], directly come from the experiment. The way of ε_{c0}^* determination is shown in the insert to Fig. 4 and explained below. Thus, the only adjustable parameter remains the coefficient A_4. To find A_4 we calculate $\sigma'(\varepsilon)$ using Eq. (15) and fit the experiment in a range of 3D AL fluctuations near T_c where $ln\sigma'(ln\varepsilon)$ is the linear function of ε with a slope $\lambda = -1/2$ (Eq. (11)). Solving Eq. (15) for the pseudogap $\Delta^*(T)$ one can readily obtain [27]

$$\Delta^*(T) = T \ln \frac{e^2 A_4 \left(1 - \frac{T}{T^*}\right)}{\sigma'(T) 16 \hbar \, \xi_c(0) \sqrt{2 \varepsilon_{c0}^*} \, \sinh(2\varepsilon / \varepsilon_{c0}^*)}, \qquad (16)$$

where $\sigma'(T)$ is the experimentally measured excess conductivity over the whole temperature interval from T^* down to T_c^{mf}.

3. Experimental results with respect to the Local Pair model

3.1. YBCO films with different oxygen concentration

Within proposed LP model the FLC and PG in YPrBCO films [94], in slightly doped $HoBa_2Cu_3O_{7-\delta}$ single-crystals [95], and evev in FeAs-based superconductor $SmFeAsO_{0.85}$ with T_c =55 K [20] (see division 3.4 below) were successfully studied for the first time. As a result, convincing set of self-consistent and reproducible results was obtained which has to corroborate the LP model approach. But the basic results have been obtained from the analysis of the resistivity data for the set of four YBCO films with different oxygen concentration [2, 27, 67, 68]. The films were fabricated at Max Plank Institute (MPI) in Stuttgart by pulse laser deposition technique [96]. All samples were the well structured c-oriented epitaxial YBCO films, as it was confirmed by studying the correspondent x-ray and Raman spectra [93]. The sample F1 (T_c=87.4 K) close to optimally doped systems, the sample F6 (T_c=54.2 K) which represents weakly doped HTS systems, and the samples F3 and F4 with T_c near 80 K were investigated to obtain the required information. Fig. 3 displays the temperature dependencies of the longitudinal resistivity $\rho_{xx}(T) = \rho(T)$ of the experimental films with parameters shown in Table I, where d_0 is the sample thickness.

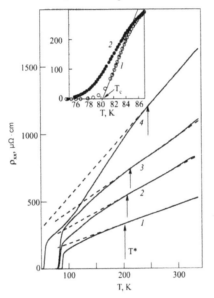

Figure 3. Temperature dependencies of ρ_{xx} for the samples F1(1), F3 (2), F4 (3), and F6 (4). Inset: $\rho_{xx}(T)$ for sample F4 (T_c=80.3 K) in zero magnetic field (1) and for **H** = 0.6 T (2).

The inset shows $\rho(T)$ for the sample F4 (T_c=80.3 K) in zero magnetic field **H**=0 (curve 1), showing how T_c was determined, and at **H** = 0.6T (curve 2), confirming the phase uniformity of the samples. Comparing the results with similar dependencies obtained for single crystals [97], the oxygen index of our samples can be estimated as follows: $\Delta y = (7 - y) \cong 6.85$ (F1), $\Delta y \cong 6.8$ (F3), $\Delta y \cong 6.78$ (F4), and $\Delta y \cong 6.5$ (F6). The resistivity parameters of the films are listed in Table I.

To simplify our discussion a little bit, I will consider only the sample F1, as an example. The similar results were obtained for all other YBCO films being studied and compared with those obtained for $HoBa_2Cu_3O_{7-\delta}$ [95] and $SmFeAsO_{0.85}$ [20]. Besides, I will consider mainly the basic aspects of the PG analysis only and touch on the FLC results as far as is necessary. To look for more details, one can see Refs.[2, 13, 20, 27, 67, 68, 94, 95].

Sample	d_0 ()	T_c (K)	T_c^{mf} (K)	$\rho(100K)$ ($\mu\Omega cm$)	$\rho(300K)$ ($\mu\Omega cm$)	T^* (K)	$\xi_c(0)$ ()
F1	1050	87.4	88.46	148	476	203	1.65
F3	850	81.4	84.55	237	760	213	1.75
F4	850	80.3	83.4	386	1125	218	1.78
F6	650	54.2	55.88	364	1460	245	2.64

Table 1. The parameters of the YBCO films with different oxygen concentration (sample F1–F6).

3.2. Pseudogap in YBCO films with different oxygen concentration

We proceed from the fact that the excess conductivity $\sigma'(T)$ arises as a result of the formation of paired fermions (local pairs) at temperatures $T_c < T < T^*$ [2, 6, 13, 23, 26, 47]. It is believed that formation of such pairs gives rise to their real binding energy, ε_b, which the quantity Δ^* characterizes [27]. As a result, the density of states of the normal excitations in this energy range decreases [57], which is referred to as the appearance of a pseudogap in the excitation spectrum [59, 61, 84, 98].

Since the PG is not measured directly in our experiments, the problem reduces to determining $\Delta^*(T)$ from the experimental dependence $\sigma'(T)$ and comparing it with those obtained from Eq. (10). To perform the analysis, the excess conductivity of every studied YBCO film, measured in the whole temperature interval from T^* down to T_c, was treated in the framework of the LP model using Eq. (15) and Eq. (16) [2, 13, 27]. Aside from the parameters, which directly come from experiment (Tables I), we substitute into Eq. (15) the values of $\Delta^*(T_c)$. Here, by analogy to the superconducting state, $\Delta^*(T_c)$ is the value of the PG in the limit $T \rightarrow T_c$. As it was shown in Ref. [15], $\Delta^*(T_c) \cong \Delta(0)$ and, correspondingly, Δ^* satisfies the condition $2\Delta^* \sim k_B T_c$, as it was demonstrated in Ref's. [99–101]. The conclusion has subsequently been confirmed by the tunneling experiments in Bi compounds [52]. To justify the values of $\Delta^*(T_c)$ used in our analysis we applied an approach proposed in Ref. [88] in which, however, the fluctuation contributions into $\sigma'(T)$ are neglected. But a definite advantage of their representation of the experimental data in the coordinates $ln\sigma'$ versus $(1/T)$ is the fact that the rectilinear part of the resulting plot has turned out to be very sensitive to the value of $\Delta^*(T_c)$, which makes it possible to adjust the value chosen for this parameter. As expected, matching is achieved for values of $\Delta^*(T_c)$ which are determined by the relation $2\Delta^*(T_c)/k_B T_c \cong 5$ [99–101]. For the sample F4 $\Delta^*(T_c) \approx 190K$, i.e., $2\Delta^*(T_c)/(k_B T_c) \simeq 4.75$ [2, 27].

The curve constructed for F1 using Eq. (15) with the parameters ξ_{c0}^* =0.233, $\xi_c(0) = 1.65$ Å, T_c^{mf} =88.46 K, T^*=203 K, $\Delta^*(T_c)$=218 K, and $A_4 = 20$ is labeled with the number 4 in Fig. 4. As one can see from the figure, the equation (15) describes well the experimental curve (thick solid line marked by I) over the whole temperature interval from T^* down to T_c. Similar results were obtained for the all other films studied.

Curve numbered 3 in the figure reproduces result of Ref. [102] in which, however, both the dynamics of pair-creation and pair-breaking above T_c were neglected. But, in accordance to our knowledge, the authors of the Ref. [102] were the first who have paid attention to the experimental fact that in YBCO compounds the reciprocal of the excess conductivity $\sigma'^{-1}(T)$ is an exponential function of ε in a certain temperature range above T_{c0}. Correspondingly, the adjustable coefficient A_3 [102] is chosen so that the computed curve matches experiment in the

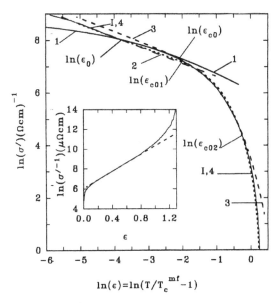

Figure 4. $\sigma'(T)$ in the coordinates $ln\sigma'$ versus $ln\varepsilon$ (solid curve **I**) for sample F1 for T from T_c to T^* in comparison with theory: curve 1-Maki-Thompsom contribution; 2-Aslamasov-Larkin contribution; 3 - theory [102]; 4 - Eq. (15) (short dashed segment). Inset: $\ln \sigma'^{-1}$ versus ε (solid line); dashed line - extrapolation of the rectilinear section [2, 27].

region of exponential behavior of $\sigma'^{-1}(T)$ (usually from $T_{c01} \approx 100\,K$ up to $T_{c02} \approx 150\,K$). It is clear that $ln\sigma'^{-1}$ is to be the linear function of ε, as shown in the insert of Fig. 4. The advantage of this approach is that it enables us to determine the parameter ε_{c0}^* which is reciprocal of the slope α of this linear function: $\varepsilon_{c0}^* = 1/\alpha$ [2, 102].

The same experimental dependencies of $\sigma'(T)$, as shown in Fig. 4, were obtained for all studied compounds, including $HoBa_2Cu_3O_{7-\delta}$ [95] and $SmFeAsO_{0.85}$ [20], suggesting the similar local pairs behavior in different HTS's. The fact enabled us to analyze the FLC and PG in $HoBa_2Cu_3O_{7-\delta}$ single crystals [95] and in $SmFeAsO_{0.85}$ FeAs-based superconductor [20] also in terms of the LP model, as will be discussed in the next divisions. Note, the complete coincidence of the given by Eq. (15) theoretical curve and the data (Fig. 4) is not necessary. We fit experiment by the theory to determine mostly the coefficient A_4 [2, 27], and coincidence in the 3D fluctuation region near T_c is only important, as discussed above. Nevertheless, the good coincidence of both theoretical and experimental curves obtained for all studied compounds [20, 67, 94, 95] means, in turn, that substituting into Eq. (16) the experimentally measured values of the $\sigma'(T)$ with the corresponding set of parameters, we should obtain a result which reflects the real behavior of $\Delta^*(T)$ quite closely in the experimental samples. The values of $\Delta^*(T)$ calculated using Eq. (16) for all YBCO films with the similar sets of parameters, as designated above for the film F1, are shown in Fig. 5 which actually displays our principal result. Indeed, despite the rather different $\Delta^*(T_c)$ and all other parameters, very similar $\Delta^*(T)$ behavior is observed for all studied films. The main common feature of every plot is a maximum of $\Delta^*(T)$ observed at the same $T_{max} \approx 130$ K. The important point here is

Superconductors: Experimental Aspects

that the coherence length $\xi_{ab}(T_{max})$ was found to be the same for every studied film, namely, $\xi_{ab}(T_{max}) \approx 18\text{Å}$ [2, 27].

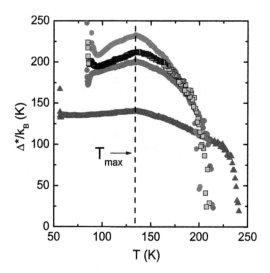

Figure 5. Dependencies of Δ^*/k_B on T calculated by Eq. (16) for samples F1 (upper curve, circles), F3 (squares), F4 (dots) and F6 (low curve, triangles). $T_{max} = T_{pair} \approx 130\,K$.

Let us discuss the obtained results (Fig. 5) now. Above 130 K $\xi_{ab}(T)$ is very small ($\xi_{ab}(T^*) \approx 13$ Å), whereas the coupling energy ε_b is very strong. It is just the condition for the formation of the SBB [6, 24–26]. It was found [27] that over the temperature interval $T_{max} = T_{pair} < T < T^*$ every experimental $\Delta^*(T)/\Delta^*(T_{max})$ curve shown in Fig. 5 can be fitted by the Babaev-Kleinert (BK) theory [87] in the BEC limit (low n_f) in which the SBB have to form [6, 24–26, 85, 87]. (See also Fig. 13 as an example). The finding has to confirm the presence of the local pairs in the films at $T \leq T^*$ which are supposed to exist at these temperatures just in the form of SBB. As SBB do not interact with one another, the local pairs demonstrate no SC (collective) behavior in this temperature interval. It has subsequently been shown to be consistent with the tunneling experiments in Bi compounds [52] in which the SC tunneling features are smeared out above T_{pair}. Thus, above T_{max} (Fig. 5), which is also called T_{pair} in accordance with, e.g., Ref. [80], it is the so-called non-superconducting part of the PG.

But the pairs have already formed and exist in the sample even in this temperature range, this is a point. There a few evidences as for paired fermions existence in temperature interval from T^* down to T_{pair} [2]. Firstly, from studying the nuclear magnetic resonance (NMR) in weakly doped Y123 systems [57], the anomalous decrease of the Knight shift K(T) was observed on cooling just at $T \leq T^*$ (Fig. 6). In Landau theory [75] $K \sim \rho_n(\varepsilon) \equiv \rho_F$, where $\rho_n(\varepsilon)$ is the energy dependence of the density of Fermi states in the normal phase, which in classical superconductors actually remains constant in the whole temperature range of the normal phase existence, i.e., approximately down to T_c. Thus, the decrease of K(T) directly

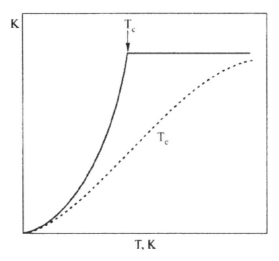

Figure 6. Temperature dependence of the Knight shift K(T) in classical superconductors (solid line) and in HTS's (dashed line) [[57]].

points out at the decrease of density of states most likely because of the local pair formation. Secondly, in measurements of the Hall effect noticeable enhancement of the Hall coefficient $R_H(T) \sim 1/e\,n_f$ was found on cooling just below T^* [2, 81, 97]. Obviously, the increase of R_H directly points out at the decrease of the normal charge-carrier density n_f. It is believed that it is because of the fact that any part of electrons transform into the local pairs. Thirdly, studying the behavior of the Nb-YBCO point contacts in high-frequency fields, we observed the pair-breaking effect of the microwave power up to $T^* \approx 230\,K$ [103]. It is clear, the observation of the pair-breaking effect naturally implies the existence of the paired fermions in the sample at such high temperatures. And, finally, very recently the polar Kerr effect (PKE) in Bi2212 was reported which also appears just below $T^* \approx 130\,K$ [90]. To be observable the PKE evidently acquires the presence of two different kinds of the quasi-particles in the sample. It is believed that such particles are most likely the normal electrons and local pairs.

Let us turn back to Fig. 5. With decreasing temperature below T_{max}, $\xi_{ab}(T)$ continues to increase whereas $\varepsilon_b(T)$ becomes smaller. Ultimately, at $T \leq T_{max} = T_{pair} \approx 130\,K$, where $\xi_{ab}(T) > 18\text{Å}$, the local pairs begin to overlap and acquire the possibility to interact. Besides, they do can be destroyed by the thermal fluctuations now, i.e. transform into fluctuating Cooper pairs, as mentioned above. The SC (collective) behavior of the local pairs in this temperature region is distinctly observed in many experiments [52, 67, 78–80], as discussed in details in div. 2.2. Eventually, the direct imaging of the local pair SC clusters persistence up to approximately 140 K in Bi compounds is recently reported in Ref. [16] (Fig.2). Thus, below T_{max} it is the SC part of the pseudogap. Moreover, we consider $\xi_{ab}(T_{max}) \approx 18\text{Å}$ to be the critical size of the local pair, at least in the YBCO [2, 27]. In fact, the local pairs behave like SBB when $\xi_{ab}(T) < 18\text{Å}$, and transform into fluctuating Cooper pairs when $\xi_{ab}(T) > 18\text{Å}$ below T_{max}, thus resulting in the BEC-BCS transition [2, 27]. That is why, I will call T_{max} as T_{pair} now, as it becomes of common occurrence in the literature [80].

3.3. Fluctuation conductivity and pseudogap in $HoBa_2Cu_3O_{7-\delta}$ single crystals under pressure

To proceed with further understanding of physical nature of HTS's, it seemed to be rather interesting to compare above results for YBCO films, where no noticeable magnetism is expected, with any other HTS's in which magnetic subsystem could play significant role. One of a such system is $HoBa_2Cu_3O_{7-\delta}$ which has magnetic moment $\mu_{eff} \approx 9.7\mu_B$, due to magnetic moment of pure Ho which is $\mu_{eff} \approx 10.6\mu_B$ [95]. Thus, when Y is substituted by Ho a qualitatively another behavior of the system is expected because of Ho magnetic properties. Besides, in HTS compounds of the $ReBa_2Cu_3O_{7=\delta}$ (Re = Y, Ho, $Dy \cdots$) type with reduced n_f the specific non-equilibrium state can be realized under change of temperature [104] or pressure [105]. In our experiments effect of hydrostatic pressure up to 5 kbar on the fluctuation conductivity $\sigma'(T)$ and pseudogap $\Delta^*(T)$ of weakly doped high-T_c single crystals $HoBa_2Cu_3O_{7-\delta}$ (HoBCO) with $T_c \approx 62$ K and oxygen index 7-$\delta \approx 6.65$ was studied [95]. The comparison of results with those obtained for $YBa_2Cu_3O_{7-\delta}$ (div. 3.2) and FeAs-based superconductor $SmFeAsO_{0.85}$ (div. 3.4) is to shed more light on the role of magnetic subsystem in the HTS's.

Measurements were carried out with current flowing in parallel to the twin boundaries when their influence on the charge carriers scattering is minimized [95]. Obtained results are analyzed within the Local Pair model [27]. The FLC and PG analysis of the sample under pressure is mainly discussed, and all results are summarized in the end. As usual, below the PG temperature $T^* \gg T_c$ resistivity of HoBCO, $\rho(T)$, deviates down from linearity resulting in the appearance of the excess conductivity $\sigma'(T) = \sigma(T) - \sigma_N(T)$ (Eq.(1)). Resulting $\ln\sigma'$ as a function of $\ln\varepsilon$ near T_c is displayed in Fig. 7.

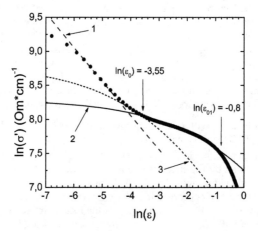

Figure 7. $\ln\sigma'$ vs $\ln\varepsilon$ for $P = 4.8$ kbar (dots). Curve 1 - AL term; 2 - MT term with $d_1 = 2.95$Å; 3 - MT term with $d = 11.68$Å. Arrows designate T_0 and T_{01}.

As expected in the LP model, above the mean field critical temperature $T_c^{mf} \approx 65.4$ K and up to $T_0 \approx 67.3$ K ($\ln\varepsilon_0 \approx -3.55$), experimental $ln\sigma'(ln\varepsilon)$ can be well extrapolated by the

3D fluctuation term of the AL theory (Eq.(11)) (Fig. 7, dashed line 1). Above T_0, up to $T_{01} \approx 95$ K (ln $\varepsilon_{01} \approx -0.8$), the experimental data are well extrapolated by the MT fluctuation contribution (Eq. (3)) (Fig. 7, curve2) of the HL theory [72]

Here I would like to emphasize that curve 2 in Fig. 7 is plotted with $d_1 = 2.95$Å, where d_1 is the distance between conducting CuO_2 planes in HoBaCuO [106], and with $\tau_\phi (100K) \beta = (0.665 \pm 0.002) \times 10^{-13}$s which is defined by a formula $\tau_\phi \beta T = A\varepsilon_{01}$ (Eq. (14)). Accordingly, d_1 corresponds to $T_{01} \approx 95$ K (ln $\varepsilon_{01} \approx -0.8$) marked by the right arrow in the figure. At $T \leq T_{01}$, $\xi_c(T) \geq d_1$ is believed to couple the CuO_2 planes by Josephson interaction, and 2D FLC has to appear [67, 77]. This scenario of the FLC appearance reminds that observed for the $SmFeAsO_{0.85}$ sample (see div. 3.4), and found $ln\sigma'(ln\varepsilon)$ is close to that shown in Fig. 10. If we choose $d \approx 11.7$Å = c, which is a dimension of the unit sell along the c-axis, as in the case of YBCO films [67], the theory noticeably deviates from experiment (curve 3). Strictly speaking, this curve reflects the temperature dependence of the FLC being close to that obtained for the YBCO film F1 shown in Fig. 4.

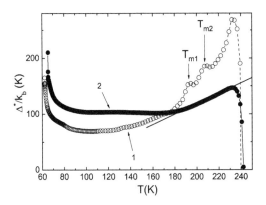

Figure 8. Δ^*/k_B as a function of T. Curve 1 - P = 0; 2 - P = 4.8 kbar. Arrows designate maxima temperatures at P = 0: T_{m1} and T_{m2}.

Thus, compare results one may conclude that HoBaCuO obviously demonstrates the enhanced 2D fluctuation contribution, compared to YBCO, and behaves in a good many ways like the $SmFeAsO_{0.85}$ (Fig. 10), most likely due to influence of its magnetic subsystem. Nevertheless, at $T = T_0 \approx 67.3$ K (ln $\varepsilon_0 \approx -3.55$), at which $\xi_c(T) = d$, the habitual dimensional 2D-3D (AL - MT) crossover is distinctly seen on the plot. Below T_0 $\xi_c(T) > c$, and 3D fluctuation behavior is realized in which the fluctuating pairs interact in the whole volume of the sample [67]. It should be emphasized that, as well as it was found for the $SmFeAsO_{0.85}$ [20], at both T_{01} and T_0 (arrows in Fig. 7) $\xi_c(0) = d_1\varepsilon_{01}^{1/2} = d\varepsilon_0^{1/2} = (1.98 \pm 0.005)$Å. The finding is believed to confirm the validity of our analysis.

The similar $ln\sigma'$ vs $ln\varepsilon$ was also obtained for $P = 0$ ($T_c = 61.4$ K, $\rho(100K) = 200\mu\Omega$cm, $T_0 = 63.4$ K, $T_{01} = 79.4$ K, $\tau_\phi(100K)\beta = 1.13 \cdot 10^{-13}$s and $d_1 = 3.2$Å). Thus, the pressure leads to increase in T_0 and T_{01}, i.e. increases the interval of 3D and especially 2D fluctuations. Most likely, it is due to decrease of the phase stratification of the single crystal under pressure

[104, 105]. The value of the C-factor also points out at this effect: $C_{3D} = 1.1$ and $= 0.53$ at $P = 4.8$ kbar and $P = 0$ kbar, respectively. The closer C_{3D} to 1.0, the more homogeneous is the sample structure [2, 67].

The experimental dependence of $\sigma'(T)$ in the whole temperature interval from T^* and down to T_c^{mf} turned out to be very close to that, shown in Fig. 4. The finding enables us to calculate $\Delta^*(T)$ using Eqs. (15) and (16). Resulting temperature dependence of the PG is shown in Fig. 8. In the figure curve 1 is plotted for $P = 0$ using Eq.(16) with the following parameters derived from experiment: $\varepsilon_{c0}^* = 0.88$, $\xi_{c(0)} = 1.57$Å, $T_c^{mf} = 62.3$ K, $T^* = 242$ K, $A_4 = 4.95$ and $\Delta^*(T)/k = 155$ K. Curve 2 is plotted for $P = 4.8$ kbar with $\varepsilon_{c0}^* = 0.67$, $\xi_c(0) = 1.98$Å, $T_c^{mf} = 65.4$ K, $T^* = 243$ K, and $A_4 = 18$, $\Delta^*(T)/k = 160$ K, respectively. In this case $T_c = 64$ K and $\rho(100)$K $= 172\mu\Omega$cm.

When $P = 0$, $\Delta^*(T)$ exhibits two unexpected maxima at $T_{m1} \approx 195$ K and at $T_{m2} \approx 210$ K (Fig. 8, curve 1), most likely because of two-phase stratification of the single crystal [104, 105]. Pressure-induced enhancement of the rising diffusion processes is assumed to cause a redistribution of labile oxygen from the low-temperature phase poor in charge carriers to the high-temperature one [105]. It results in the disappearance of both peaks and linear $\Delta^*(T)$ dependence with a positive slope at high T (Fig. 8, curve 2). Simultaneously, the sample resistivity, ρ, noticeably decreases, whereas T_c and $\xi_c(0)$ somewhat increase. Note that the maxima of $\Delta^*(T)$ and initial values of ρ, T_c and $\xi_c(0)$ are restored when the pressure is removed suggesting the assumption. Thus, both $\sigma'(T)$ and $\Delta^*(T)$ are markedly different from those obtained for the YBCO films [27, 67]. They resemble similar curves obtained for FeAs-based superconductor $SmFeAsO_{0.85}$ [20] (see div. 3.4). The result can be explained by the influence of paramagnetism in $HoBa_2Cu_3O_{7-\delta}$ [95, 104, 105]. Additionally, the linear drop of the PG, found for both $HoBa_2Cu_3O_{7-\delta}$ [95] and $SmFeAsO_{0.85}$ [20], is believed to reflect the influence of weak magnetic fluctuations in such compounds [21, 107, 108], as will be discussed in more details in the next division.

3.4. Fluctuation conductivity and pseudogap in $SmFeASO_{0.85}$

Despite of a huge amount of papers devoted to FeAs-based superconductors, there is an evident lack of the FLC and PG studies in these compounds [20]. This state of affairs is most likely due to the extreme complexity of the electronic configuration of the FeAs-based compounds, where the strong influence of magnetism, being changed with doping, is observed [17–19]. It is well known that upon electron or hole doping with F substitution at the O site [17, 113], or with oxygen vacancies [17, 18] all properties of parent RFeAsO compounds drastically change and evident antiferromagnetic (AFM) order has to disappear [17–19]. However, recent results [107–110] point toward an important role of low-energy spin fluctuations which emerge on doping away from the parent antiferromagnetic state which is of a spin-density wave (SDW) type [107, 108, 110]. Thus, below $T_S \sim 150K$ the AFM fluctuations, being likely of spin wave type, are believed to noticeably affect the properties of doped $RFeAsO_{1-x}F_x$ systems (Fig. 9). As shown by many studies [108–110] the static magnetism persists well into the SC regime of ferropnictides. In $SmFeAsO_{1-x}$ strongly disordered but static magnetism and superconductivity both are found to coexist in the wide range of doping, and prominent low-energy spin fluctuations are observed up to the highest achievable doping levels where T_c is maximal [21, 107, 108].

The interplay between superconductivity and magnetism has been a long-standing fascinating problem [111, 112], and relation between the SDW and SC order is a central topic in the current research on the FeAs-based high-T_c superconductors. However, the clear nature of the complex interplay between magnetism and superconductivity in FeAs-based HTS's is still rather controversial [112]. As a result, rather complicated phase diagrams for different FeAs-based high-T_c systems [17, 109, 110], and especially for $SmFeAsO_{1-x}$ [113–116] (Fig. 9) are reported.

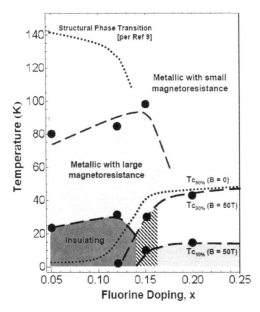

Figure 9. Phase diagram of $SmFeAsO_{1-x}F_x$ [[113]].

Naturally, rather peculiar normal state behavior of the system upon T diminution is expected in this case [109, 110, 113], when x is of the order of 0.15, as it is in our sample [20].

3.4.1. Experimental details

To shed more light on the problem, analysis of the FLC and PG was carried out within the LP model using results of resistivity measurements of $SmFeAsO_{0.85}$ polycrystal ($T_c = 55$ K) performed on fully computerized set up [20]. The width of the resistive transition into superconducting (SC) state is $\Delta T \leq 2K$ suggesting good phase and structural uniformity of the sample. In accordance with the LP model approach the resistivity curve above $T^* \sim 170K$ is extrapolated by the straight line to get $\sigma'(T) = \sigma(T) - \sigma_N(T)$ using Eq.(1).

The crossing of measured $\sigma'^{-2}(T)$ with T-axis denotes the mean-field critical temperature $T_c^{mf} \cong 57K$. This is the usual way of T_c^{mf} determination within the LP model [2, 67]. Resulting $\ln\sigma'$ versus $\ln\varepsilon$ is displayed in Fig. 10 in the temperature interval relatively close to T_c (dots) and compared with the HL theory [72] in the clean limit (curves 1-3). As expected, up to $T_0 \approx$

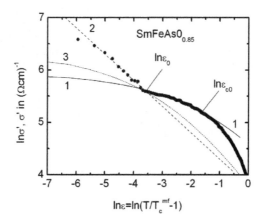

Figure 10. $ln\sigma'$ as a function of $ln\varepsilon$ near T_c (dots) in comparison with the HL theory: 1-MT contribution (d = 3.05Å); 2 - 3D AL contribution; 3-MT contribution (d = 8.495Å).

58.5 K ($ln\varepsilon_0$ =-3.6) the data are well extrapolated by the AL fluctuation contribution (Eq. 11) for any 3D system (straight line 2 in the figure). As mentioned above, this 3D fluctuation behavior was found to be typical for all without exception HTS compounds [2, 20, 67, 80, 95]. Accordingly, above T_0, up to $T_{c0} \approx 69K$ ($ln\varepsilon_{c0} \cong -1.55$), this is 2D MT term (Eq. (3)) of the HL theory (Fig. 10, curve 1) which dominates well above T_c in the 2D fluctuation region [2, 67], as discussed in details in div. 2.2. As before, $\xi_c(0) = d\varepsilon_0^{1/2}$ is the coherence length along the c-axis, i.e. perpendicular to the conducting planes, and d is the distance between conducting layers in $SmFeAsO_{1-x}$. Thus, expected MT-AL (2D-3D) crossover is clearly seen in Fig. 10 at $ln\varepsilon_0$ =-3.6. The fact enables us to determine T_0 (Eq. (7)) and therefore the ε_0 (Eq. (7, 9)) with adequate accuracy. Now, proceeding in the usual way, i.e., using Eqs. (12, 14), and set d = 8.495Å, which is a dimension of the $SmFeAsO_{0.85}$ unit sell along the c-axis [114], $\xi_c(0) = (1.4 \pm 0.005)$Å, and $\tau_\phi(100K)\beta = (11 \pm 0.03) \times 10^{-13}$ s are derived from the experiment. As it is seen from the figure, Eq. (11) with measured value of $\xi_c(0)$ and the scaling factor $C_3D = 0.083$ describes the data fairly well just above T_c^{mf} (Fig. 10, dashed line 2) suggesting 3D fluctuation behavior of $SmFeAsO_{0.85}$ near T_c.

Till now the FeAs-based superconductor behaves like the YBCO films. However, the discrepancy appears when MT contribution is analyzed (Fig. 10, curve 3). Indeed, substituting the measured values of $\xi_c(0)$ and $\tau_\phi(100K)\beta$ into Eq. (3) we obtain curve 3 which looks like that found for YBCO films but apparently does not match the data. The finding suggests that our choice of d is very likely wrong in this case. To proceed with the analysis we have to suppose that $SmFeAsO_{1-x}$ becomes quasi-two-dimensional just when $\xi_c(T)$, getting rise with temperature diminution, becomes equal to $d_1 = 3.05$ Å at $T = T_{co} \approx 69K$ ($ln\varepsilon_{c0} \cong -1.55$). It is worth to emphasize that $(3.1 \div 3.0)$Å is the distance between As layers in conducting $As - Fe - As$ planes in $SmFeAsO_{1-x}$ [18, 114]. Below this temperature $\xi_c(T)$ is believed to couple the As layers in the planes by Josephson interaction [95]. This approach gives the same value of $\xi_c(0) = d\,\varepsilon_{c0}^{1/2} = (1.4 \pm 0.005)$Å as calculated above using the usual crossover temperature T_0. We think this fact is to confirm the supposition. At the same time the phase relaxation time changes noticeably and amounts $\tau_\phi(100K)\,\beta = (1.41 \pm 0.03) \times 10^{-13}$s in this case. Substitution of measured values of $\xi_c(0)$ and this new $\tau_\phi(100K)\,\beta$ into Eq. (3) enables us to fit the experimental data by the MT term just up to T_{c0} which is about 15 K above T_c (Fig. 10, curve 1). The result suggests enhanced 2D MT fluctuations in $SmFeAsO_{1-}$ compared to YBCO films. Recall, the similar enhanced 2D MT fluctuations are found for $HoBa_2Cu_3O_{7-\delta}$ [95] (div. 3.3)

3.4.2. Pseudogap analysis

Somehow surprisingly, $ln\sigma'$ versus $ln\varepsilon$ measured for the $SmFeAsO_{0.85}$ in the whole temperature range from T^* down to T_c^{mf} was found to be very close to that of YBCO films (Fig. 4). Even the exponential dependence of the reciprocal of the excess conductivity $\sigma'^{-1}(T)$ still occurs between $ln\varepsilon_{c0}$ and $ln\varepsilon_{c02}$ ((69 - 100)K). As a result, the function $ln\sigma'^{-1}$ versus ε appears to be linear in this temperature range (similar to that shown in insert in Fig. 4), and its slope $1/\alpha$ still determines the parameter ε_{c0}^* used in the PG analysis [20, 27]. Taking obtained results into account, one can draw a conclusion that the LP model approach can be applied to analyze the temperature dependence of the PG. To proceed with the PG analysis the value of the coefficient A_4 must be found. As before, to determine A_4 we fit the experimental $\sigma'(\varepsilon)$ by the theoretical curve (Eq. (15)) in the region of 3D fluctuations near T_c [20, 27]. All other parameters directly come from experiment, as discussed above. The theoretical $\sigma'(\varepsilon)$ curve is constructed using Eq. (15) with the reasonable set of experimentally measured parameters: ξ_{c0}^* = 0.616, $\xi(0) = 1.4$Å, $T_c^{mf} = 56.99$ K, $T^* = 175K$, $A_4 = 1.98$ and optionally chosen $\Delta^*(T_c)/k_B = $ 160 K and found to describe the experimental data well in the whole temperature interval from T^* down to T_c^{mf} [95]. The only exception is the 2D MT region where relatively small deviation down from experiment, negligible in the case of YBCO films [67], is observed. It is due to enhanced MT fluctuation contribution in the 2D fluctuation region, as mentioned above. Compare results with those obtained for magnetic superconductor $Dy_{0.6}Y_{0.4}Rh_{3.85}Ru_{0.15}B_4$ [117] we assume this enhancement to be the consequence of weak magnetic fluctuations in $SaFeAsO_{1-x}$ [107–110], as will be discussed below.

Now we have almost all parameters to calculate the temperature dependence of PG using Eq.(16) except the value of $\Delta^*(T_c)$ which actually is responsible for the A_4. The problem is, the value of $\Delta^*(T_c)$ and in turn the ratio $2\Delta^*(T_c)/k_B T_c$ in Fe-based superconductors remain uncertain. Reported in the literature values for $\Delta(0)$ and $2\Delta(0)/T_c$ in $SmFeAsO_{1-x}$ range from $2\Delta(0) \approx 37$ meV ($2\Delta/T_c \sim 8$, strong-coupling limit) obtained in measurements of far-infrared permittivity [17], down to $\Delta = (8 - 8.5)$ meV ($2\Delta/T_c \sim (3.55 - 3.8)$) measured by a scanning tunneling spectroscopy [118, 119] which is very close to standard value 3.52 of the BCS theory (weak-coupling limit). It is believed at present that $SmFeAsO_{1-x}$ has two superconducting gaps $\Delta_1(0) \approx 6.7$ meV and $\Delta_2(0)$ which ranges from $\approx 15\,meV$ up to $\approx 21\,meV$ [120]. Besides we still think that $\Delta^*(T_c) \approx \Delta(0)$ [20], as is was justified in Ref. [15]. To feel more flexible, four curves are finally plotted in Fig. 11 with $\Delta^*(T_c)/k_B = 160$ K ($2\Delta^*(T_c)/T_c \cong 5.82$), = 140 K ($2\Delta^*(T_c)/T_c \cong 5.0$), 120 K ($2\Delta^*(T_c)/T_c \cong 4.36$) and = 100 K ($2\Delta^*(T_c)/T_c \cong 3.63$) from top to bottom, respectively. Naturally, different values of coefficients A_4 correspond to each curve, whereas the other above parameters remain unchangeable.

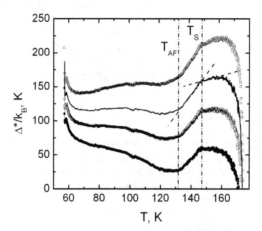

Figure 11. Δ^*/k_B versus T dependencies for $SmFeAsO_{0.85}$ with four different values of $\Delta^*(T_c)/k_B$ (see the text).

Very unexpected $\Delta^*(T)$ behavior is observed (Fig. 11). The most striking result is a sharp drop of $\Delta^*(T)$ at $T_s \approx 147$ K, as clearly illustrates the curve with $\Delta^*(T_c)/k_B$ =140 K plotted without symbols. Usually T_s is treated as a temperature at which a structural tetragonal-orthorhombic transition occurs in parent SmFeAsO. In the undoped FeAs compounds it is also expected to be a transition to SDW ordering regime [17–19]. Below T_S the pseudogap $\Delta^*(T)$ drops linearly down to $T_{AFM} \approx 133$ K (Fig. 11), which is attributed to the AFM ordering of the Fe spins in a parent SmFeAsO compounds [17, 121]. Below T_{AFM} the slop of the $\Delta^*(T)$ curves apparently depends on the value of $\Delta^*(T_c)$ [20]. Strictly speaking it is difficult to say at present is $T_{AFM} = T_N$ of the whole system or not because the AFM ordering of Sm spins occurs at $\approx 5K$ only [17, 121].

Found $\Delta^*(T)$ behavior is believed to be explained in terms of the theory by Machida, Nokura and Matsubara (MNM) [111] developed for AFM superconductors in which the AFM ordering with a wave vector Q may coexist with the superconductivity. This assumption was supported in, eg., the theory by Chi and Nagi [122]. In the MNM theory the effect of the AFM molecular field $h_Q(T)$ ($|h_Q| \ll \varepsilon_F$) on the Cooper pairing was studied. It was shown, that below T_N the BCS coupling parameter $\Delta(T)$ is reduced by a factor $[1 - const \cdot |h_Q(T)|/\varepsilon_F]$ due to the formation of energy gaps of SDW on the Fermi surface along Q. As a result the effective attractive interaction $\breve{g} N(0)$ or, equivalently, the density of states at the Fermi energy ε_F is diminished by the periodic molecular field that is

$$\breve{g} N(0) = g N(0) [1 - \alpha m(T)]. \qquad (17)$$

Here $m(T)$ is the normalized sublattice magnetization of the antiferromagnetic state and α is a changeable parameter of the theory. Between T_c and T_N ($T_c > T_N$ is assumed) the order parameter is that of the BCS theory. Since below T_N the magnetization $m(T)$ becomes nonvanishing, $\breve{g} N(0)$ is weakened that results in turn in a sudden linear drop of $\Delta(T)$

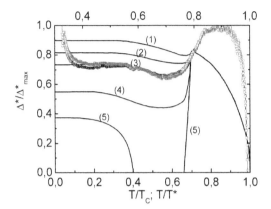

Figure 12. $\Delta^*(T)/\Delta_{max}$ in $SmFeAsO_{0.85}$: (red $\circ$) - $\Delta^*(T_c)/k_B$=130 K; ($\circ$) - 135 K. Solid curves $\Delta(T)/\Delta_{max}$ correspond to MNM theory with the different $\alpha \sim 1/[\mathfrak{g}\,N(0)]$: (1) - $\alpha = 0.1$, (2) - $\alpha = 0.2$, (3) - $\alpha = 0.3$, (4) - $\alpha = 0.6$, (5) - $\alpha = 1.0$; $T_N/T_c = 0.7$ [[111]].

immediately below T_N. As $m(T)$ saturates at lower temperatures, $\Delta(T)$ gradually recovers its value with increasing the superconducting condensation energy (Fig. 12, solid curves). This additional magnetization $m(T)$ was also shown to explain the anomaly in the upper critical field H_{c2} just below T_N observed in studying of RMo_6S_8 (R = Gd, Tb, and Dy) [111]. However, predicted by the theory decrease of $\Delta(T)$ at $T \leq T_N$ was only recently observed in AFM superconductor $ErNi_2B_2C$ with $T_c \approx 11$ K and $T_N \approx 6$ K, below which the SDW ordering is believed to occur in the system [123]. The result evidently supports the prediction of the MNM theory. Our results are found to be in a qualitatively agreement with the MNM theory as shown in Fig. 12, where the data for $\Delta^*(T_c)/k_B = 130$ K (red $\circ$) and $\Delta^*(T_c)/k_B = 135$ K ($\circ$) are compared with the MNM theory (solid lines).

The curves are scaled at $T/T_c = 0.7$ and demonstrate rather good agreement with the theory below $T/T_c = 0.7$. Note, that the upper scale is T/T^*. Both shown $\Delta^*(T)$ dependencies suggest the issue that just $\Delta^*(T_c)/k_B = 133$ K, which is just T_{AFM}, would provide the best fit with the theory. Above $T/T_c = 0.7$ the data evidently deviate from the BCS theory. It seems to be reasonable seeing $SmFeAsO_{0.15}$ as well as any other ferropnictides is not a BCS superconductor.

It is important to emphasize that in our case we observe the particularities of $\Delta^*(T)$ in the PG state, i. e. well above T_c but just at $T \approx T_s$, below which the SDW ordering in parent SmFeAsO should occur. However, it seems to be somehow surprising as no SDW ordering in optimally doped $SmFeAsO_{0.15}$ is expected. The found very specific $\Delta^*(T)$ can be understood taking mentioned above weak AFM fluctuations (low-energy spin fluctuations), which should exist in the system, into account. At the singular temperature T_s these fluctuations are believed to enhance AFM in the system likely in the form of SDW. After that, in accordance with the MNM theory scenario, the SDW has to suppress the order parameter of the local pairs as shown by our results. Thus, the results support the existence both weak AFM fluctuations and the paired fermions (local pairs) in the PG region, which order parameter is apparently suppressed by these fluctuations [20].

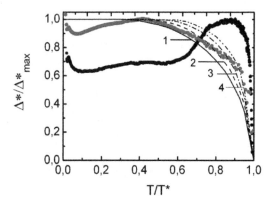

Figure 13. $\Delta^*(T)/\Delta_{max}^*$ in $SmFeAsO_{0.85}$ with $\Delta^*(T_c)/k_B = 160$ K ($\bullet$) and in YBCO film with $T_c = 87.4$ K (blue $\bullet$) [27] as a function of T/T^* (T/T_c in the case of the theory). Curves $1 \cdots 4$ correspond to BK theory with different $x_0 = \mu/\Delta(0)$: 1 - $x_0 = 10.0$ (BSC limit), 2 - $x_0 = -2.0$, 3- $x_0 = -5.0$, 4 - $x_0 = -10.0$ (BEC limit) [[87]].

To confirm the conclusion, relation $\Delta^*(T)/\Delta_{max}^*$ as a function of T/T^* (T/T_c in the case of the theory) is plotted in Fig. 13. Black dots represent the studied $SmFeAsO_{0.85}$ with $\Delta^*(Tc)/k_B$ = 160 K, which corresponds to the strongly coupled limit ($2\Delta^*(T_c)/T_c \cong 5.82$). Blue dots display the data for YBCO film F1 [27]. The solid and dashed curves display the results of the Babaev-Kleinert (BK) theory [87] developed for the SC systems with different charge carrier density n_f. For the different curves the different theoretical parameter $x_0 = \mu/\Delta(0)$ is used, where μ is the chemical potential. Curve 1, with $x_0 = +10$, gives the BCS limit. Curve 4 with x_0 = -10 represents the strongly coupled BEC limit, which corresponds to the systems with low n_f in which the SBB must exist [23, 25, 85, 87]. As well as in the YBCO film, the $\Delta^*(T)/\Delta_{max}^*$ in $SmFeAsO_{0.85}$ evidently corresponds to the BEC limit (Fig. 13) suggesting the local pairs presence in the FeAs-based superconductor. Below $T/T^* \approx 0.4$ both experimental curves demonstrate the very similar temperature behavior suggesting the BEC-BCS transition from local pairs to the fluctuating Cooper pairs to be also present in FeAS-based superconductors [20]. But, naturally, no drop of $\Delta^*(T)$ is observed for the YBCO film (Fig. 13, blue dots) at high temperatures, as no antiferromagnetism is expected in this case. This fact can be considered as an additional evidence of the AFM nature of the $\Delta^*(T)$ linear reduction below T_s in $SmFeAsO_{0.85}$ found in our experiment.

It has to be emphasized that recently reported phase diagrams [113–116] (Fig. 9) apparently take into account a complexity of magnetic subsystem in $SmFeAsO_{1-x}F_x$ and are in much more better agreement with our experimental results. But it has to be also noted that we study the $SmFeAsO_{1-x}$ system whereas the phase diagrams are mainly reported for the $SmFeAsO_{1-x}F_x$ compounds. Is there any substantial difference between the both compounds has yet to be determined. Evidently more experimental results are required to clarify the question.

3.5. Angle-resolved photoemission measurements of the energy pseudogap of high-T_c $(Bi, Pb)_2(Sr, La)_2CuO_{6+\delta}$ superconductors: A model evidence of local pairs

Taking all above consideration into account, it can be concluded that the FLC and PG description in terms of local pairs gives a set of reasonable and self-consistent results. However, to justify the conclusion it would be appropriate to test the LP model approah using independent results of other research groups who have measured straightforwardly the PG or any other related effects. But for a long time there was a lack of indispensable data.

Fortunately, analysis of the pseudogap in $(Bi, Pb)_2(Sr, La)_2CuO_{6+\delta}$ (Bi2201) single-crystals with various T_c's by means of ARPES spectra study was recently reported [80]. The study of Bi2201 allows avoid the complications resulting from the bilayer splitting and strong antinodal bosonic mode coupling inherent to Bi2212 and Bi2223 [90, 91]. Symmetrized energy distribution curves (EDCs) were found to demonstrate the opening of the pseudogap on cooling below T^*. It was shown that T^*, obtained from the resistivity measurements, agrees well with one determined from the ARPES data using a single spectral peak criterion [80]. Finally, from the ARPES experiments information about the temperature dependence of the loss of the spectral weight close to the Fermi level, $W(E_F)$, was derived [80]. $W(E_F)$ versus T measured for optimally doped OP35K Bi2201 (T_c=35 K, T^* = 160 K) turned out to be rather unexpected, as shown in Fig. 14a taken from Ref. [80]. Above T^* the $W(E_F)$ is nonlinear function of T. But below T^*, over the temperature range from T^* to $T_{pair} = (110 \pm 5)$ K (Fig. 14a), the $W(E_F)(T)$ decreases linearly which is considered as a characteristic behavior of the "proper" PG state [80]. However, no assumption as for physical nature of this linearity as well as for existence of the paired fermions in the PG region is proposed. Below T_{pair} the $W(E_F)$ vs T noticeably deviates down from the linearity (Fig. 14a). The deviation suggests the onset of another state of the system, which likely arises from the pairing of electrons, since the $W(E_F)(T)$ associated with this state smoothly evolves through T_c (Fig. 14a).

To compare results and justify our Local Pair model, the ρ_{ab} vs T of the OP35K Bi2201, reported in Ref. [80], was studied within the LP model [13]. The $\Delta^*(T)$ was calculated by Eq. (16) with the following reasonable set of parameters: T_c = 35 K, T_c^{mf} = 36,9 K, T^* =160 K, $\xi_c(0)$= 2.0Å, ε_{c0}^* = 0.89, and A_4 = 59. $\sigma'(T)$ is the experimentally measured excess conductivity derived from the resistivity data using Eq. (1). Resulting $\Delta^*(T)/\Delta_{max}^*$ is plotted in Fig. 14b (green dots). As expected, the shape of the $\Delta^*(T)$ curve is similar to that found for YBCO films (Fig. 5). Besides, the maximum of $\Delta^*(T)/\Delta_{max}^*$ at $T_{max} \approx 100K$ actually coincides with the change of the $W(E_F)(T)$ slop at T_{pair}, measured by ARPES, which seems to be reasonable. In fact, in accordance with our logic, T_{max} is just the temperature which divides the PG region on SC and non-SC parts depending on the local pair state, as described above. Recall, that above T_{max} the local pairs are expected to be in the form of SBB. Most likely just the specific properties of SBB cause the linear $W(E_F)(T)$ over this temperature region (Fig. 14a). The two following facts are believed to confirm the conclusion. Firstly, when SBB disappear above T^*, the linearity disappears too. Secondly, below T_{max}, or below T_{pair} in terms of Ref. [80], the SBB have to transform into fluctuating Cooper pairs giving rise to the SC (collective) properties of the system. This argumentation coincides with the conclusion of Ref. [80] as for SC part of the pseudogap below T_{pair}. As SBB are now also absent, the linearity of $W(E_F)(T)$ again disappears. Thus, we consider the $\Delta^*(T)$, calculated within our Local Pair model (Fig. 14b), to be in a good agreement with the temperature dependence of the loss of the spectral weight

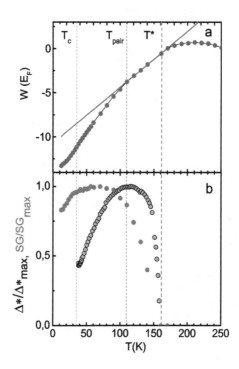

Figure 14. a. Spectral weight $W(E_F)$ vs T (blue dots) for OP35K Bi2201. The solid line is the guidance for eyes only [[80]]. b. Pseudogap $\Delta^*(T)/\Delta^*_{max}$ (green dots) and spectral gap SG/SG_{max} (red dots) [[80]] as the functions of temperature for the same sample.

$W(E_F)$ (Fig. 14a) obtained from the ARPES experiments performed on the same sample. In this way, the results of ARPES experiments reported in Ref. [80] are believed to confirm our conclusion as for existence of the local pairs in HTS's, at least in Bi2201 compounds.

Also plotted in Fig 14b is the normalized spectral gap ($SG(T)$) (red dots) equals to the energy of the spectral peaks of EDCs measured by ARPES [80]. Important in this case is that $SG(T)$ smoothly evolves through both T_{pair} and T_c. The fact is believed to confirm assumed in the LP model the local pair existence above T_{pair}. Despite the evident similarity there are, however, at least two differences between the curves shown in Fig. 14b. First, there is no direct correlation between the $SG(T)$ and the $W(E_F)(T)$ (Fig. 14 a, b). Why the maximum of $SG(T)$ is shifted toward low temperatures compared to T_{pair}, has yet to be understood. The second difference is the absolute value of the SG compared to the pseudogap. The spectral gap has $SG_{max} \approx$ 40 meV and $SG(T_c) \approx 38$ meV [80]. It gives $2SG(T_c)/k_BT_c \approx 26$ which is apparently too high. The PG values are $\Delta^*_{max} \approx 16.5$ meV and $\Delta^*(T_c) \approx 6.96$ meV, respectively. It gives $2\Delta^*(T_c)/k_BT_c \approx 6.4$ which is a common value for the Bi compounds [124] with respect to relatively low T_c in this case. Thus, there is an unexpected lack of coincidence between the SG and PG.

4. Conclusion

The Chapter presents a detailed consideration of the LP model developed to study the PG in HTS's. In accordance with the model the local pairs have to be the most likely candidate for the PG formation. At high temperatures ($T_{pair} < T \leq T^*$) we believe the local pairs to be in the form of SBB which satisfy the BEC theory (non-SC part of a PG). Below T_{pair} the local pairs have to change their state from the SBB into fluctuating Cooper pairs which satisfy the BCS theory (SC part of a PG). Thus, with decrease of temperature there must be a transition from BEC to BCS state [2, 27]. The possibility of such a transition is considered to be one of the basic physical principals of the high-T_c superconductivity. The transition was predicted theoretically in Ref. [23, 24, 74] and experimentally observed in our experiments [2, 20, 27, 95].

A key test for our consideration is the comparison of the $\Delta^*(T)$, calculated within the LP model, with the temperature dependence of the loss of the spectral weight close to the Fermi level $W(E_F)(T)$, measured by ARPES for the same sample [80]. The resulting $\Delta^*(T)$ is found to be in a good agreement with the $W(E_F)(T)$ obtained for OP35K Bi2201 (Fig. 14). It allows us to explain reasonably the $W(E_F)(T)$ dependence, both above and below T_{pair}, in terms of local pairs.

The obtained results are also in agreement with the conclusions of Ref's. [16, 90, 91] as for SC and non-SC parts of the PG in Bi systems. Besides, formation of the local pairs is also believed to explain the rise of the polar Kerr effect and response of the time-resolved reflectivity, both observed for Bi systems just below T^* [90]. While, the Nernst effect [16], which is likely due to the SC properties of the local pairs, is observed only below T_{pair}, or below T_{max} in terms of our model. All the facts have to support the local pair existence in HTS's at $T \leq T^*$, Thus, we may conclude, that on the basis of the developed LP model the self-consistent picture of the PG formation in HTS's is obtained. At the same time the issue concerning the pairing mechanism in HTS's still remains controversial [22].

Acknowledgments

The author is grateful to V. M. Loktev for valuable discussions and to T. Kondo for critical remarks.

Author details

Andrei L. Solovjov

B. I. Verkin Institute for Low Temperature Physics and Engineering of National Academy of Science of Ukraine, Lenin ave. 47, Kharkov 61103, Ukraine

5. References

[1] J. G. Bednorz , K. A. Mueller. "Possible High - T_c Superconductivity in the Ba-La-Cu-O System", *Z. Phys. B. - Condensed Matter*, vol. 64, pp. 189-193, 1986.

[2] A. L. Solovjov, V. M. Dmitriev, "Fluctuation Conductivity and Pseudogap in YBCO High-Temperature Superconductors", *Low Temp. Phys.* vol. 35, pp. 169-197, 2009.

[3] C. P. Slichter, " Experimental Evidence for Spin Fluctuations in High Temperature Superconductors", *Strongly Correlated Electronic Materials*, K. S. Bedell, Z. Wang, D. Meltzer, A. Balatsky, and E. Abrahams, Ed., Redwood City: Addison-Wesley, 1994, p. 427-479.

[4] Yu. A. Izyumov, "Strongly Correlated Electrons: t-J-Model", *Physics Uspekhi*, vol. 167, pp. 465-497, 1997.

[5] J. M. Tranquada, G. D. Gu, M. Hücker et al. (13 auth.), "Evidence for Unusual Superconducting Correlations Coexisting with Stripe Order in $La_{1.875}Ba_{0.125}CuO_4$", *Phys. Rev. B*, vol. 78, pp. 174529(1-4), 2008.

[6] V. M. Loktev, "Particularities of Superconductivity in 2D Metals: Transition from Cooper Pairing to the Local One", *Low Temp. Phys.*, vol. 22, pp. 488-491, 1996.

[7] A. A. Pashitskii, "Low-frequency Fluctuations of the Charge Density and High-Temperature Superconductivity in Metallic-Oxides Compounds", *Low Temp. Phys.* vol. 21, pp. 763-777, 1995; and *Low Temp. Phys.*, vol. 21, pp. 837-883, 1995.

[8] T. Timusk and B. Statt, "The Pseudogap in High-Temperature Superconductors: an Experimental Survey", *Rep. Prog. Phys.*, vol. 62, pp. 161-222, 1999.

[9] Y. Iye, *Phys. Properties of High-Temperature Superconductors*, D. M. Ginsberg, Ed., vol.3, Singapore: World Scientfic, 1992, pp. 285-361.

[10] V. V. Eremenko, V. N. Samovarov, V. N. Svishchev, V. L. Vakula, M. Yu. Libin, and S. A. Uyutnov, "Observation of the Habbard and Covalent Correlations in Spectral Characteristics of $YBa_2Cu_3O_{6+x}$ Films", *Low Temp. Phys.*, vol. 26, pp. 541-556, 2000.

[11] B. P. Stojkovic, D. Pines, "Theory of the Longitudinal and Hall Conductivities of the Cuprate Superconductors", *Phys. Rev. B*, vol. 55, pp. 8576-8595, 1997.

[12] E. G. Maksimov, "The Problem of High-Temperature Superconductivity. The Update Situation", *Physics Uspekhi*, vol. 43, pp. 965-993, 2000.

[13] A. L. Solovjov, M. A. Tkachenko, "Pseudogap and Local Pairs in High-T_c Cuprate Superconductors", *arXiv:1112.3812v1 [cond-mat.supr-con]*.

[14] C.W. Chu, L. Gao, F. Chen, Z. J. Huang, R. L. Meng,and Y. Y. Xue, "Superconductivity Above 150 K in $HgBa_2Ca_2Cu_3O_{8+\delta}$ at High Presures", *Nature*, vol. 365, pp. 323-324, 1993.

[15] J. Stajic, A. Iyengar, K. Levin, B. R. Boyce, and T. R. Lemberger," Cuprate Pseudogap: Competing Oder Parameters or Precursor Superconductivity", *Phys. Rev. B*, vol. 68, pp. 024520-024529, 2003.

[16] A. Yazdani, "Visualizing Pair Formation on the Atomic Scale and the Search for the Mchanism of Superconductivity in high-T_c cuprates", *J. Phys.: Condens. Matter*, vol. 21, pp. 164214-164219, 2009.

[17] M V. Sadovskii, *Physics Uspekhi*, "High-Temperature Superconductivity in Layered Iron Compounds", vol. 51, pp. 1201-1271, 2008.

[18] A. L. Ivanonskii, "New High-Temperature Superconductors Based on Rare Earth and Transition Metal Oxyarsenides and Related Phases: Synthesis, Properties and Simulation", *Physics Uspekhi*, vol 51, pp. 1229- 1306, 2008.

[19] Y. A. Izyumov and E. Z. Kurmaev, *Physics Uspekhi*, vol. 51, pp. 1261- 2008.

[20] A. L. Solovjov, V. N. Svetlov, V. B. Stepanov, S. L. Sidorov, V. Yu. Tarenkov, A. I. D'yachenko, and A.B.Agafonov, "Possibility of Local Pair Existence in Optimally Doped

$SmFeAsO_{1-x}$ in Pseudogap Regime", *arXiv:1012.1252v [cond-mat, supr-con]*, 2010; *Low Temp. Phys.*, vol. 37, pp. 557-601, 2011.

[21] I. I. Mazin, "Superconductivity gets an iron boost", *Nature*, vol. 464, pp. 11-13, 2010.

[22] V.M. Loktev, V.M. Turkowsky, "On the Theory of a Pseudogap State in Low Dimensional Superconductors with Anisotropic Order Parameter", *Rep. National Academ. Sci. Ukr.*, vol. 1, pp. 70-75, 2012.

[23] Ñ.A.R. Sa de Melo, M. Randeria, and J. R. Engelbrecht, "Crossover from BCS to Bose Superconductivity: Transition Temperature and Time-Dependent Ginzburg-Landau Theory", *Phys. Rev. Lett.*, vol. 71, pp. 3202-3205, 1993.

[24] R. Haussmann, "Properties of a Fermi Liquid at the Superfluid Transition in the Crossover Region Between BCS Superconductivity and Bose-Einstein Condensation", *Phys. Rev. B*, vol. 49, pp. 12975-12983, 1994.

[25] J. R. Engelbrecht, M. Randeria, and C. A. R. Sa de Melo,"BCS to Bose Crossover: Broken Symmetry State", *Phys. Rev. B*, vol. 55, pp. 15153-15156, 1997.

[26] O. Tchernyshyov, "Noninteracting Cooper Pairs Inside a Pseudogap", *Phys. Rev. B*, vol. 56, pp. 3372-3380, 1997.

[27] A. L. Solovjov and V. M. Dmitriev, "Resistive studies of the pseudogap in YBaCuO films with consideration of the transition from BCS to Bose-Einstein condensation", *Low Temp. Phys.*, vol. 32, pp. 99-108, 2006.

[28] V. J. Emery and S. A. Kivelson, "Importance of Phase Fluctuations in Superconductors with Small Superfluid Density", *Nature*, vol. 374, pp.434-437, 1995.

[29] P. W. Anderson, "The Theory of Superconductivity in High-T_c Cuprates", *Phys. Rev. Lett.*, vol. 67, pp. 2092, 1991.

[30] A. S. Alexandrov and N. F. Mott, *High Temperature Superconductors and Other Superfluids*, London: Taylor and Francis, 1994.

[31] J. T. Devreese, *Polarons*, (Encyclopedia of Applied Physics), G.L. Trigg, Ed., vol. 14, Dordrecht: VCH Publishers, 1996, pp. 383-414.

[32] J. Bonča and S. A. Trugman, " Bipolarons in the Extended Holstein Hubburd Model", *Phys. Rev. B*, vol. 64, pp. 094507(1-4), 2001.

[33] J. Bardeen, L. N. Cooper, and J. R. Schrieffer, "Theory of Superconductivity", *Phys. Rev.*, vol. 108, pp. 1175-1204, 1957.

[34] Ponomarev Ya. G., Tsokur E. B., Sudakova M. V., S. N. Tchesnokov et.al., "Evidence for Strong ElectronŰPhonon Interaction from Inelastic Tunneling of Cooper Pairs in c-Direction in $Bi_2Sr_2CaCu_2O_8$ Break Junctions", Solid State Commun. Vol. 111, pp. 513-518, 1999.

[35] K. H. Bennemann and J. B. Katterson, Ed.,*The Physics of Superconductors. Conventional and High-T_c Superconductors*, vol. 1, New York: Springer, 2003.

[36] P. W. Anderson, *The Theory of Superconductivity in the High T_c Cuprates*, Princeton: Princeton University Press, 1997.

[37] L. Benfatto and S. G. Sharapov, "Optical-Conductivity Sum Rule in Cuprates and Unconventional Charge Density Waves: a Short Review", *Low Temp. Phys.*, vol. 32, pp. 533-547, 2006.

[38] A. M. Gabovich, A. I. Voitenko, T. Ekino, Mai Suan Li, H. Szymczak, and M Pękała, "Competition of Superconductivity and Charge Density Waves in Cuprates: Recent Evidence and Interpretation", *Advances in Condensed Matter Physics*, vol. 2010, ID 681070, 40 pp. 2010.

[39] A. Mills and H. Monien, "Antiferromagnetic Correlations and Nuclear Magnetic Relaxation in High-T_c Superconductors: A Critical Reexamination", *Phys. Rev. B*, vol. 45, pp. 5059-5076, 1992.

[40] T. Sakai and Y. Takahashi, "Pseudogap Induced by Antiferromagnetic Spin Correlation in High-Temperature Superconductors", *J. Phys. Soc. Jpn.*, vol. 70, pp. 272-277, 2001.

[41] I. Eremin and D. Manske, " Excitations in Layered Cuprates: a Fermi-Liquid Approach", *Low Temp. Phys.*, vol. 32, pp. 519-532, 2006.

[42] J. R. Schrieffer, X. G. Wen, and S. C. Zhang, "Dynamic Spin Fluctuations and the Bag Mechanism of High-T_c Superconductivity", *Phys. Rev. B*, vol. 39, pp. 11663-11679, 1990.

[43] A. Kampf and J. R. Schrieffier, "Pseudogaps and the Spin-Bag Approach to High-T_c Superconductivity", *Phys. Rev. B*, vol. 41, pp. 6399-6408, 1990.

[44] J. Ruvalds, "Theoretical Prospects for High-Temperature Superconductors", *Semicond. Sci. Technol.*, vol. 9, pp. 905-926, 1996.

[45] J. F. Annet, N. Goldenfeld, A. J. Leggett, "Transport Properties of High-Tc Cuprates", D.M. Ginsberg, Ed., *Physics Properties of High-Temperature Superconductors*, vol. 5, Singapore: World Scientific, 1995. pp. 375-462.

[46] R. Zeyher and M. L. Kulič,"Renormalization of the Electron-Phonon Interaction by Strong Electronic Correlations in High-T_c superconductors", *Phys. Rev. B*, vol. 53, 2850-2862, 1996.

[47] A. V. Chubukov and J. Schmalian, "Superconductivity due to Massless Boson Exchange in the Strong-Coupling Limit". *Phys. Rev. B.*, vol. 72, pp. 174520 (1-14), 2006.

[48] V. M. Krasnov, A. Yurgens, D. Winkler, P. Delsing, and T Claeson, "Evidence for Coexistence of the Superconducting Gap and the Pseudogap in Bi-2212 from Intrinsic Tunneling Spectroscopy", *Phys. Rev. Letters*, vol. 84, p. 5860-5863, 2000.

[49] R. A. Klemm, "Striking similarities between the pseudogap phenomena in cuprates and in layered organic and dichalcogenide superconductors", *Physica C*, Vol. 341-348, pp. 839-847, 2000.

[50] T. Ekino, Y. Sezaki, H. Fujii, "Features of the Energy Gap Above T_c in $Bi_2Sr_2CaCu_2O_{8+\delta}$ as Seen by Break-Junction Tunneling", *Phys. Rev. B*, vol. 60, pp. 6916-6919, 1999.

[51] M. Suzuki and T. Watanabe, "Discriminating the Superconducting Gap from the Pseudogap in $Bi_2Sr_2CaCu_2O_{8+\delta}$ by Interlayer Tunneling Spectroscopy", *Phys. Rev. Lett.*, vol. 85, pp. 4787-4790, 2000.

[52] Y. Yamada, K. Anagawa, T. Shibauchi, T. Fujii, T. Watanabe, A. Matsuda, Sand M. Suzuki, "Interlayer Tunneling Spectroscopy and Doping Dependent Energy-Gap Structure of the Trilayer Superconductor $Bi_2Sr_2Ca_2Cu_3O_{10+\delta}$. *Phys. Rev. B*, vol. 68, pp. 054533(1-11), 2003.

[53] T. Ekino, A. M. Gabovich, Mai Suan Li, M. Pękała, H. Szymczak, and A. I. Voitenko, "Analysis of the Pseudogap-Related Structure in Tunneling Spectra of Superconducting $Bi_2Sr_2CaCu_2O_{8+\delta}$ Revealed by the Break-Junction Technique",*Phys. Rev. B*, vol. 76, pp180503(R)(1-4), 2007.

[54] M. C. Boyer, W. D. Wise, K. Chatterjee et al. (8 auth.), "Imaging the Two Gaps of the High-Temperature Superconductor $Bi_2Sr_2CuO_{6+x}$, *Nature Physics*, vol. 3, pp. 802-806, (2007)

[55] A. A. Abrikosov,"Properties of the Pseudogap Phase in High-T_c Superconductors", *Phys. Rev. B*, vol. 64, pp. 104521(1-10, 2001.

[56] J. L. Tallon and J. W. Loram, "The Doping Dependence of T^* what is the Real High-T_c Phase Diagram?", *Physica C*, vol. 349, pp. 53-68, 2001.

[57] H. Alloul, T. Ohno, and P. Mendels,"89Y NMR Evidence for a Fermi-Liquid Behavior in $YBa_2Cu_3O_{6+x}$". *Phys. Rev. Lett.*. vol. 63, pp. 1700-1703, 1989.

[58] H. Yasuoka, T. Imai, and T. Shimizu, *NMR and NQR in Highly Correlated Metallic and Superconducting Cu Oxides*, H. Fukuyama, S. Maekawa, and A. P. Malozemoff, Ed., (Strong Correlation and Superconductivity), Berlin: Springer, 1989, p. 254-261.

[59] B. Bucher, P. Steiner, J. Karpinski, E. Kaldis, and P. Wachter, "Influence of the Spin Gap on the Normal State Transport in $YBa_2Cu_3O_4$", *Phys. Rev. Lett.*, vol. 76, pp. 2012-2015, 1993.

[60] E. A. Pashitski and V. I. Pentegov, " To the Mechanism of SC and PG State Appearance in High-T_c Superconductors", *Ukr. Fiz. Zh.*, vol.50, pp. A77-A81, 2005.

[61] G. Deutscher, "Superconducting Gap and Pseudogap", *Low Temp. Phys.*, vol. 32, pp. 566-571, 2006.

[62] A. M. Gabovich and A. I. Voitenko, "Model for the Coexistence of d-Wave Superconducting and Charge-Density-Wave Order Parameters in High-Temperature Cuprate Superconductors", *Phys. Rev. B*, vol. 80, 224501 (1-9), 2009.

[63] T. Ekino, A. M. Gabovich, Mai Suan Li, M. Pękała, H. Szymczak and A. I. Voitenko, "d-Wave Superconductivity and s-Wave Charge Density Waves: Coexistence between Order Parameters of Different Origin and Symmetry", *Symmetry*, vol. 3, pp. 699-749, 2011.

[64] P. W. Anderson, "The Resonating Valence Bond State in La_2CuO_4 and Superconductivity", *Science*, vol. 233, pp. 1196-1198, 1987.

[65] P. W. Anderson and Z. Zou, "Normal" Tunneling and "Normal" Transport: Diagnostics for the Resonating-Valence-Bond State". *Phys. Rev. Lett.*, vol.60, pp. 132-135, 1988.

[66] R. K. Nkum and W. R. Datars, "Fluctuation-Enhanced Conductivity in the Sb-Doped Bi-Pb-Sr-Ca-Cu-O Superconducting Systems", *Phys. Rev. B*, vol. 44, pp. 12516-12520, 1991.

[67] A. L. Solovjov, H.-U. Habermeier, and T. Haage,"Fluctuation Conductivity in $YBa_2Cu_3O_{7-y}$ films with Different Oxygen Content. I. Optimally and Lightly Doped YBCO Films", *Low Temp. Phys.*, vol. 28, pp. 17-24, 2002.

[68] A. L. Solovjov, H.-U. Habermeier, and T. Haage, "Fluctuation Conductivity in $YBa_2Cu_3O_{7-y}$ Films with Different Oxygen Content. II. YBCO Films with $T_c \approx 80\,K$", *Low Temp. Phys.*, vol. 28, pp. 99-108, 2002.

[69] L. G. Aslamazov and A. L. Larkin, "The Influence of Fluctuation Pairing of Electrons on the Conductivity of the Normal Metal", *Phys. Lett.*, vol. 26A, pp. 238-239, 1968.

[70] K. Maki, "The Critical Fluctuation of the Order Parameter in Type-II Superconductors", *Prog. Theor. Phys.*, vol.39, pp. 897-906, 1968.

[71] R. S. Thompson, "Microwave Flux Flow and Fluctuation Resistance of Dirty Type-II Superconductors", *Phys. Rev. B.*, vol. 1, pp. 327-333, 1970.

[72] S. Hikami, A.I. Larkin, "Magnetoresistance of High Temperature Superconductors", *Mod. Phys. Lett. B*, vol.2, pp. 693-698, 1988.

[73] J. B. Bieri, K. Maki and R. S. Thompson, "Nonlocal Effect in Magneto-Conductivity of High-Tc Superconductors", *Phys. Rev. B*, vol. 44, pp. 4709 - 4711, 1991.

[74] V. L. Ginzburg, L. D. Landau, " On the Theory of Superconductivity", *JETP*, vol. 20, pp. 1064-1082, 1950.

[75] E. M. Lifshitz and L. P. Pitaevski, *Statistical Physics*, vol. 2, Moscow: Nauka, 1978.

[76] W. E. Lawrence and S. Doniach, "Theory of Layer Structure Superconductors", in *Proceedings of the 12-th International Conference on Low Temperature Physics*, Kioto, 1971, pp. 361-362.

[77] Y. B. Xie, "Superconducting Fluctuations in the High-Temperature Superconductors: Theory of the dc Resistivity in the Normal State", *Phys. Rev. B*, vol. 46, pp. 13997-14000, 1992.

[78] K. Kawabata, S. Tsukui, Y. Shono, O. Mishikami et al., "Detection of a Coherent Boson Current in the Normal State of a High-Temperature Superconductor $YBa_2Cu_3O_y$ Film Patterned to Micrometer-Size Rings", *Phys. Rev. B*, vol. 58, pp. 2458- 2461, 1998.

[79] Ch. Renner, B. Revaz, J.-Y. Genoud, K. Kadowaki, and Q. Fischer, "Pseudogap Precursor of the Superconducting Gap in Under- and Overdoped $Bi_2Sr_2CaCu_2O_{8+\delta}$", *Phys. Rev. Lett.*, vol. 80, pp. 149-152, 1998.

[80] T. Kondo, Y. Hamaya, A. D. Palczewski1, T. Takeuchi, J. S.Wen, Z. J. Xu, G. Gu, J. Schmalian and A. Kaminski, "Disentangling Cooper-pair Formation Above the Transition Temperature from the Pseudogap State in the Cuprates", *Nature Phys.*, vol. 7, pp. 21-25, 2011.

[81] A. S. Alexandrov, V. N. Zavaritsky, and S. Dzhumanov, "Hall Effect and Resistivity in Underdoped Cuprates", *Phys. Rev. B*, vol. 69, 052505(1-4), 2004.

[82] A. C. Bódi, R. Laiho, E. Lähderanta, "Boson-Fermion Fluctuation Dominated Model of the Superconducting Transition in Hole-Underdoped Cuprates", *Physica C*, Vol. 411, pp. 107-113, 2004.

[83] K. Morawetz, B. Schmidt, M. Schreiber, and P. Lipavsky, "Enhancement of Fermion Pairing due to the Presence of Resonant Cavities", *Phys. Rev. B*, vol. 72, pp. 174504 (1-5), 2005.

[84] V. M. Loktev, "Spectra and Pseudogap Phenomena in High-Temperature Superconductors", *Ukr. Fiz. Zh. Oglyadi*, vol. 1, pp. 10-48, 2004.

[85] V. N. Bogomolov, " Superconductivity of High- and Low-Temperature Superconductors as Bose-Einstein Condensation. (Diluted Metals)", *Lett. JETF*. vol, 33, p. 30-37, 2007.

[86] V. M. Loktev and V. M. Turkowski, "Doping-Dependent Superconducting Properties of Two-Dimensional Metals with Different Types of Interparticle Coupling", *Low Temp. Phys.*, vol. 30, pp. 179-191, 2004.

[87] E. Babaev, H. Kleinert, "Nonperturbative XY-model Approach to Strong Coupling Superconductivity in Two and Three Dimensions", *Phys. Rev. B*, vol. 59, pp. 12083-12089, 1999.

[88] D. D. Prokof'ev, M. P. Volkov, and Yu. A. Bojkov,"The Amount and Temperature Dependence of Pseudogap in YBCO Obtained from Resistivity Measurements", *Fiz. Tverd. Tela*, vol. 45, pp. 1168-1176, 2003.

[89] V. P. Gusynin, V. M. Loktev, and S. G. Sharapov, Phase Diagram of 2D Metal System with Variable Number of Charge Carriers", *Lett. JETF*, vol. 65, pp. 170-175, 1997.

[90] Rui-Hua He, M. Hashimoto, H. Karapetyan et al. (19 auht.), "From a Single-Band Metal to a High-Temperature Superconductor via Two Thermal Phase Transitions", *Science*, vol. 331, pp. 1579-1583, 2011.

[91] K. Nakayama, T. Sato, Y.-M. Xu, Z.-H. Pan et al. "Two Pseudogaps with Dfferent Energy Scales at the Antinode of the High-temperature $Bi_2Sr_2CuO_6$ Superconductor Using Angle-resolved Photoemission Spectroscopy", *arXiv:1105.5865v [cond-mat, supr-con]* pp. 1-5, 2011.

[92] B. Oh, K. Char, A. D. Kent, M. Naito, M. R. Beasley et al., "Upper Critical Field, Fluctuation Conductivity, and Dimensionality in $YBa + 2Cu_3O_{7-x}$", Phys. Rev. B, vol. 37, pp. 7861-7864, 1988.

[93] T. Haage, J. Q. Li, B. Leibold, M. Cardona, J. Zegenhagen, H.-U. Habermeier, "Substrate-Mediated Anisotropy of Transport Properties in $YBa_2Cu_3O_{7-\delta}$", Solid State Comm., Vol.99, pp. 553-557, 1996.

[94] A. L. Solovjov, V. M. Dmitriev, "FLuctuation Conductivity and Pseudogap in $Y_{1-}PrBa_2Cu_3O_{7-y}$ films", Low Temp. Phys., vol. 32, pp. 753-760, 2006.

[95] A. L. Solovjov, M. A. Tkachenko, R. V. Vovk, Z. F. Nazyrov, and M. A. Obolenskii, "Fluctuation Conductivity and Pseudogap in Slightly Doped $HoBa_2Cu_3O_{7-\delta}$ Single Crystals under Pressure", Low Temp. Phys., vol. 37, pp. 1053-1056, 2011.

[96] H.-U. Habermeier, "Pulsed Laser Deposition - a Versatile Technique Only for High-temperature Superconductor Thin-film Deposition?", Applied Surface Science, Vol. 69.Ü pp. 204-211, 1993.

[97] T. Ito, K. Takenaka, and S. Uchida, "Systematic Deviation from T-Linear Behavior in the In-Plane Resistivity of $YBa_2Cu_3O_{7-y}$: Evidence for Dominant Spin Scattering", Phys. Rev. Lett., vol. 78, pp. 3995-3998, 1993.

[98] K. Maki, H. Won, "Pseudogap Phenomenon in High-T_c Cuprates", J. Magnetic Mat., Vol. 226-230. pp. 350-352, 2001.

[99] K. W. Wang, W. Y. Ching, "A Structural-Based Microscopic Theory on High-Temperature Cuprate Superconductors", Physica C, 2004.vol. 416, pp. 47-67, 2004.

[100] R. O. Zaitsev, "To the Peculiareties of the Electronic Mechanism of the Superconductivity", Zh. Eksp. Teor. Fiz., vol. 125, pp. 891-905 2004.

[101] V. M. Dmitriev and A. L. Solovjov, "Energy Gap and Features of HTS's Conductivity", Fiz. Nizk. Temp. (Low.Temp.Phys.), vol.16, pp. 650-653, 1990.

[102] 201B. Leridon, A. Defossez, J. Dumont, J. Lesueur, and J. P. Contour, ". Conductivity of Underdoped $YBa_2Cu_3O_{7-\delta}$: Evidence for Incoherent Pair Correlations in the Pseudogap Regime", Phys. Rev. Lett., vol. 87, pp. 197007 (1-4), 2001.

[103] V.M. Dmitriev, A. L. Solovjov, A. I. Dmitrenko, L. A. Ischenko, "Nonlinear Properties of $YBa_2Cu_3O_x$ under Microwave Irradiation in the Temperature Range 4.2 - 270 Ê", Fiz. Nizk. Temp. (Low Temp. Phys.), vol. 15, pp. 1223-1225, 1989.

[104] R. V. Vovk, M. A. Obolenskii, A. A. Zavgorodniy, I. L. Goulatis, V. I. Beletskii, A. Chroneos, "Structural Relaxation, Metal-to-Insulator Transition and Pseudo-Gap in Oxygen Deficient $Ba_2Cu_3O_{7-\delta}$ Single Crystals", Physica C, vol. 469, pp. 203-206, 2009.

[105] R. V. Vovk, Z. F. Nazyrov, M. A. Obolenskii, I. L. Goulatis, A. Chroneos, S. V. Pinto, "Phase Separation in Oxygen Deficient $Ba_2Cu_3O_{7-\delta}$ Single Crystals: Effect of High Pressure and Twin Boundaries", Phil. Mag., vol. 91, pp. 2291-2302, 2011.

[106] G. D. Chryssikos, E. I. Kamitsos, J. A. Kapoutsis et al. (8 auth.), "X-ray Diffraction and Infrared Investigation of $RBa_2Cu_3O_7$ and $R_{0.5}Pr_{0.5}Ba_2Cu_3O_7$ Compounds (R, Y and Lanthanides)", Physica C, vol. 254, pp. 44-62, 1995.

[107] A. J. Drew, Ch. Niedermayer, P. J. Baker, F. L. Pratt, S. J. Blundell et al. (16 auth.), "Coexistence of Static Magnetism and Superconductivity in $SmFeAsO_{1-x}F_x$ as Revealed by Muon Spin Rotation", Nature materials, vol. 8, pp. 310-317, 2009.

[108] S. Sanna, R. De Renzi, G. Lamura et al. (8 auth.), "Magnetic-Superconducting Phase Boundary of SmFeAsO1-xFx Studied via Muon Spin Rotation: Unified Behavior in a Pnictide Family", Phys. Rev. B, vol, 80, pp. 052503 (1-4), (2009).

[109] L.F. Zhu and B.G. Liu, "Striped Antiferromagnetism and Electronic Structures of SrFeAsF and Their Implications", *EPL*, vol. 85, pp 67009-67013, 2009.

[110] R. M. Fernandes, D. K. Pratt, W. Tian et al (13 auth), "Unconventional pairing in the iron arsenide superconductors", *Phys. Rev. B*, vol. 81, 140501(R1-R4), 2010.

[111] K. Machida, "Coexistence Problem of Magnetism and Superconductivity" *Appl. Phys. A*, vol. 35, 193-215, 1984.

[112] W. Jiao, Q. Tao, J. Bao, et al. (8 auth), "Anisotropic Superconductivity in $Eu(Fe_{0.75}Ru_{0.25})_2As_2$ Ferromagnetic Superconductor", EPL, vol. 95, pp. 67007(1-6), 2011.

[113] S. C. Riggs, J. B. Kemper, Y. Jo, Z. Stegen, L. Balicas, G. S. Boebinger, F. F. Balakirev et al. (11 auth), "Log-T Divergence and Insulator-to-Metal Crossover in the Normal State Resistivity of Fluorine Doped $SmFeAsO_{1-x}F_x$ *arXiv: Cond-mat/0806.4011.*, 2008.

[114] S. Margadonna, Y. Takabayashi, M. T. McDonald et al. (8 auth.), "Crystal Structure and Phase Transitions Across the Metal-Superconductor Boundary in the $SmFeAsO_{1-x}F_x$ $(0 \leq x \leq 0.20)$ Family *Phys. Rev. B*, vol. 79, pp, 014503(1-7), 2009.

[115] Y. Kamihara, T. Nomura, M. Hirano et al. (9 auth.) "Electronic and Magnetic Phase Diagram of Superconductors, SmFeAsO1-xFx", *New Journal of Physics*, vol.12, pp 033005(1-14), (2010).

[116] M. Tropeano, C. Fanciulli et al. (15 auth.), "Transport and Infrared Properties of SmFeAs(O1-xFx): From SDW to Superconducting Ordering", *Supercond. Sci. Technol.*, vol. 22, pp. 034004(1-7), 2009.

[117] A. V. Terekhov, M. A. Tkachenko, E. P. Khlybov, A. Zaleski, and A. L. Solovjov, "Fluctuation Conductivity, Pseudogap, Magnetic and Transport Properties of a Magnetic Superconductor $Dy_0.6Y_0.4Rh_3.85Ru_0.15B_4$", unpublished.

[118] O. Millo, I. Asulin, O. Yuli, et al. (8 auth.), "Scanning Tunneling Spectroscopy of $SmFeAsO_{0.85}$: Possible Evidence for d-Wave Order-Parameter Symmetry", *Phys. Rev. B*, vol. 78, pp. 092505(1-3), 2008.

[119] T. Y. Chen, Z. Tesanovic, R. H. Liu, X. H. Chen, C. L. Chien, "A BCS-like Gap in the Superconductor $SmFeAsO_{0.85}F_{0.15}$", *Nature*, vol. 453, pp. 1224-1228, 2008.

[120] D. Daghero, M. Tortello, R. S. Gonnelli, V. A. Stepanov, N. D. Zhigadlo, and J.Karpinski, "Evidence for Two-Gap Nodeless Superconductivity in $SmFeAsO_{0.8}F_{0.2}$ from Point-Contact Andreev-Reflection Spectroscopy", *arXiv:0812.1141v1 [cond-mat.supr-con]*, 2008.

[121] L. Ding, C. He, J. K. Dong, T.Wu, R. H. Liu, X. H. Chen, and S. Y. Li, "Specific Heat of the Iron-Based High-Tc Superconductor $SmO_{1-x}F_xFeAs$", *Phys. Rev. B*, vol. 77, pp. 180510R(1-4), 2008.

[122] Hongguang Chi and A. D. S. Nagi, "Transport Properties of Antiferromagnetic Superconductors", *Journal of Low Temperature Physics*, vol. 86, pp. 139-160, 1992

[123] N. L. Bobrov, V. N. Chernobay, Yu. G. Naidyuk et al. (9 auth.), "Competition of Multiband Superconducting and Magnetic Order in $ErNi_2B_2C$ Observed by Andreev Reflection", *EPL*, vol. 83, pp. 37003p (1-5), 2008.

[124] Ya. Ponomarev, M. Mikheev, M. Sudakova, S. Tchesnokov, and S. Kuzmichev, "Extended van Hove Singularity, Strong ElectronŰPhonon Interaction and Superconducting Gap in Doped Bi-2212 Single Crystals", *Phys. Status Solidi C*, vol. 6, pp. 2072-2075, 2009.

Permissions

The contributors of this book come from diverse backgrounds, making this book a truly international effort. This book will bring forth new frontiers with its revolutionizing research information and detailed analysis of the nascent developments around the world.

We would like to thank Alex Gabovich, for lending his expertise to make the book truly unique. He has played a crucial role in the development of this book. Without his invaluable contribution this book wouldn't have been possible. He has made vital efforts to compile up to date information on the varied aspects of this subject to make this book a valuable addition to the collection of many professionals and students.

This book was conceptualized with the vision of imparting up-to-date information and advanced data in this field. To ensure the same, a matchless editorial board was set up. Every individual on the board went through rigorous rounds of assessment to prove their worth. After which they invested a large part of their time researching and compiling the most relevant data for our readers. Conferences and sessions were held from time to time between the editorial board and the contributing authors to present the data in the most comprehensible form. The editorial team has worked tirelessly to provide valuable and valid information to help people across the globe.

Every chapter published in this book has been scrutinized by our experts. Their significance has been extensively debated. The topics covered herein carry significant findings which will fuel the growth of the discipline. They may even be implemented as practical applications or may be referred to as a beginning point for another development. Chapters in this book were first published by InTech; hereby published with permission under the Creative Commons Attribution License or equivalent.

The editorial board has been involved in producing this book since its inception. They have spent rigorous hours researching and exploring the diverse topics which have resulted in the successful publishing of this book. They have passed on their knowledge of decades through this book. To expedite this challenging task, the publisher supported the team at every step. A small team of assistant editors was also appointed to further simplify the editing procedure and attain best results for the readers.

Our editorial team has been hand-picked from every corner of the world. Their multi-ethnicity adds dynamic inputs to the discussions which result in innovative outcomes. These outcomes are then further discussed with the researchers and contributors who give their valuable feedback and opinion regarding the same. The feedback is then collaborated with the researches and they are edited in a comprehensive manner to aid the understanding of the subject.

Apart from the editorial board, the designing team has also invested a significant amount of their time in understanding the subject and creating the most relevant covers. They scrutinized every image to scout for the most suitable representation of the subject and create an appropriate cover for the book.

The publishing team has been involved in this book since its early stages. They were actively engaged in every process, be it collecting the data, connecting with the contributors or procuring relevant information. The team has been an ardent support to the editorial, designing and production team. Their endless efforts to recruit the best for this project, has resulted in the accomplishment of this book. They are a veteran in the field of academics and their pool of knowledge is as vast as their experience in printing. Their expertise and guidance has proved useful at every step. Their uncompromising quality standards have made this book an exceptional effort. Their encouragement from time to time has been an inspiration for everyone.

The publisher and the editorial board hope that this book will prove to be a valuable piece of knowledge for researchers, students, practitioners and scholars across the globe.

List of Contributors

Guoqing Wu
Department of Physics, University of West Florida, USA

W. Gilbert Clark
Department of Physics and Astronomy, University of California, Los Angeles, USA

A.J.S. Machado, S.T. Renosto, C.A.M. dos Santos and L.M.S. Alves
Escola de Engenharia de Lorena, Universidade de São Paulo, Lorena, SP, Brazil

Z. Fisk
Departments of Physics and Astronomy, University of California at Irvine, Irvine, USA

Chi Liang Chen and Chung-Li Dong
Institute of Physics, Academia Sinica , Nankang, Taipei, Taiwan
National Synchrotron Radiation Research Center (NSRRC), Hsinchu, Taiwan

Mitsuru Izumi
TUMSAT-OLCR, Tokyo, Japan

Jacques Noudem
CRISMAT/LUSAC-UNICAEN, Caen, France

Gunzi Saito and Yukihiro Yoshida
Faculty of Agriculture, Meijo University, Shiogamaguchi 1-501 Tempaku-ku, Nagoya, Japan

Qiuliang Wang, Zhipeng Ni and Chunyan Cui
Institute of Electrical Engineering, Chinese Academy of Sciences, China

Laureline Porcar, Patricia de Rango, Daniel Bourgault and Robert Tournier
Institut Néel/CRETA/CNRS/ University Joseph Fourier/Grenoble/France

Andrei L. Solovjov
B. I.Verkin Institute for Low Temperature Physics and Engineering of National Academy of Science of Ukraine, Lenin ave. 47, Kharkov 61103, Ukraine

Printed in the USA
CPSIA information can be obtained
at www.ICGtesting.com
JSHW011402221024
72173JS00003B/388